Product Intervention

This book covers all aspects of design thinking and its intervention as needed for developing products for the elderly. The book deals with the Universal Principles of Design and how they can be included with Technological Interventions for showcasing the product lifecycle.

The main technical features include the Anthropometry calculations, Design Thinking approach to the healthcare products, and selection of target group, which is elderly in the presented context. Further, the complete product development cycle, the choice of materials, assessment of the designs, prototyping phases and the final product are all discussed in detail. The various fabrication strategies to reduce the cost of the product, ideation stages and Feedback and Assessment right from design to product development stage are discussed.

All the Anthropometry details are included specifically for the target group chosen; its mathematical understanding, error, etc., are all discussed in the book. The ideations, rendering and conceptualization based on Cognitive Ergonomics or Hawthorne and various other effects are also discussed in detail.

Product Intervention
User-Centric Design to Fabrication

Biswajeet Mukherjee

CRC Press
Taylor & Francis Group
Boca Raton London New York

CRC Press is an imprint of the
Taylor & Francis Group, an **informa** business

First edition published 2023

by CRC Press
60 Broken Sound Parkway NW, Suite 3, Boca Raton, FL 33487-2742

and by CRC Press
4 Park Square, Milton Park, Abingdon, Oxon, OX14 4RN

CRC Press is an imprint of Taylor & Francis Group, LLC

© 2024 Biswajeet Mukherjee

ISBN: 978-1-032-39269-1 (hbk)
ISBN: 978-1-032-54060-3 (pbk)
ISBN: 978-1-003-41495-7 (ebk)

DOI: 10.1201/9781003414957

Typeset in Times New Roman
by SPi Technologies India Pvt Ltd (Straive)

Contents

Preface

Design and technology have remained a much-needed segment in most areas of technology and spheres of design. It is interesting to note that design students are facilitated to pursue case studies which are applicable to the real world; however, the technology part remains a niche area for them to work. While engineers are encouraged to execute a solution to a problem through a protocol, the designers identify the grass root problem and then try to propose various feasible solutions.

There is plenty of text available that cite that when design and technology do not go hand in hand, there is a strong likelihood of observing a catastrophe or a failure. Further, while most design books deal with the principles and applications of design, engineering or technology books deal with the formulas, protocols and concepts without including the inputs of the designers.

This book aims to bridge the gap by taking specific case studies that deal with the design to technology end of the journey of any product. Elderly health care is a prima facie for any country or family; thus, the elderly are the focus for whom the products are redesigned along with technological intervention. The various practices of Design Thinking approach and the tools of the research are also adopted and explained with reference to the example of the case study to understand the journey of a product. I am sure that the readers will feel a strong connection not only with the challenges posed in the design process but also with the technological solutions provided.

Acknowledgements

It gives me immense pleasure to thank the Science for Equity and Empowerment Division (SEED) of the Department of Science and Technology, Ministry of Science and Technology, Government of India, for its generous support. I was sponsored by a SEED Division Project entitled 'Technology Intervention in Product design for the Elderly – Case Studies on Umbrella and Stick Designs' vide the F. No. SP/YO/2/2017 dated November 1, 2017. Dr Rashmi Sharma, as the Coordinator from the SEED division, has always ensured that the best support be rendered from the funding agency. I extend my sincere gratitude to her.

A number of students have been part of the journey in the making of this book: beginning with my three students from the Design Department of IIITDM Jabalpur, namely Ms Shivangi Pande, Mr Akshay Kenjale and Mr Aditya Mathur, for the design evaluation, concepts and ideation. Ms Maitreeyi Gautami and Mr Nikhil Krishna Reddy, students of the ECE department, worked with the fabrication process. Mr P. Daniel Akhil Kumar was the appointed Junior Research Fellow of the project, and his contribution throughout the journey of the project has been unparalleled.

I also thank my PhD students, namely Dr Vinay Killamsetty, Dr Monika Chauhan, Mr Anil Rajput, Mr Pramod Kumar Gupta, Ms Manshree Mishra and my MTech student Mr Sourodipto Das for the creation of figures and sketches. I am also grateful to Dr Prabir Mukhopadhyay, Head of the Department of Design Discipline, PDPM IIITDM Jabalpur, for all the advice he has given during the journey of the project and the book.

Finally, I thank my family for their unparallel support and their patience throughout. My mother Ms Papya Mukherjee, who contributed at all levels of Design, Prototyping and Testing, deserves my indebtedness. I also thank my wife, Mrs Jayashree Mukherjee, for having been extremely supportive throughout. This book is dedicated in honour of my father, Late Shri Abhoy Kumar Mukherjee, for his teachings and the values he has instilled in me. I wish I can pass the legacy to my infant daughter Ms Aaina Mukherjee, to become a good human being.

Author

Biswajeet Mukherjee has completed his Bachelor of Technology (B. Tech) in Electronics and Communication Engineering from GGSIP University, Delhi; Master of Technology (M. Tech) in Microwave Electronics from the Department of Electronic Science, University of Delhi, and his Doctor of Philosophy (PhD) from the Department of Electrical Engineering, Indian Institute of Technology Bombay, Powai, Mumbai. He is currently working as Assistant Professor in the Department of Electronics and Communication Engineering, PDPM Indian Institute of Information Technology, Design & Manufacturing, Jabalpur.
In his career, he has completed several externally sponsored research projects from various organisations like the Science and Engineering Research Board (SERB) of the Department of Science & Technology (DST), Government of India; Science for Equity and Empowerment Division (SEED), DST; Madhya Pradesh Council of Science and Technology (MPCST), Government of Madhya Pradesh, etc.

He has successfully guided three PhD students and six are completing their PhD under his supervision. He has already guided ten master's theses as well. He is a Member of the Editorial Board of the International Journal of Applied Electromagnetics and Mechanics, published by the IOS press, the Netherlands. He was selected among the top 2% scientists of the world based on the list issued by Stanford University, USA. He is the recipient of the Best Paper Award at the International Conference on Innovative Product Design and Intelligent Manufacturing Systems (ICIPDIMS), 2019 at NIT Rourkela, India. He was awarded Hari Om Ashram Prerit Harivallabhdas Chunilal Shah Research Endowment Prize in the field of Electronics & Communication for the year 2015–2016, declared in 2017, by Sardar Patel University, Gujarat. He was awarded the INSA Visiting Scientist Fellowship for 2017–2018. He received the Young Engineer Award of the Institution of Engineers India (IEI), at Vijayawada, India, in 2016. He was a recipient of the 'Young Scientist Award' at the 31st URSI General Assembly and Scientific Symposium (GASS) in Beijing, China, August 16–23, 2014.

His research areas include Microwave and Antenna Engineering, Electromagnetics, Dielectric Resonators and Product Design in Electronics.

1 Introduction to Design and Technology

It is interesting to note that every product that we come across has some scientific principle laid as its foundation. In fact, science, which works on ground reality, is better termed as 'engineering'. Engineering is the use of scientific principles and temperament to design and fabricate machines, prototypes, structures and products, bringing them to reality.

Technology, in contrast, is the application of engineering or scientific principles to evolve the latest state of the art of know-how in engineering practices. In short, technology refers to the skills, processes and methods to achieve certain clearly defined objectives and goals. It is interesting to observe that while both engineering and technology refer to practical applications, they work in the backend of any system. In general, the various engineering streams working in the back end are shown in Figure 1.1.

The classical branches of engineering refer to civil engineering, electrical engineering and mechanical engineering. With the advent of time and the changing pace of technology, newer streams like computer science and engineering, information technology, electronics and communication engineering, electronics and electrical engineering, mechatronics, embedded systems, Internet of things, automation and production engineering and aerospace engineering were added. This is shown in the second-level hierarchy of Figure 1.1.

Some of these streams are in fact multidisciplinary in nature. For example, Embedded Systems and Internet of Things is a blend of Electronics, Computers and mechanical systems. The combination of the three streams has led to innovation and evolution. It is interesting to note the new range of products which are added to the current market by virtue of these new streams. For example, the concept of smart homes or smart devices is possible by capable backend programming of microcontrollers while accepting real-time inputs from the dynamic environment. For the said purpose, a database is needed, a programming language needs to be prioritized and the electronic components and the space in which the components need to be incubated also has to be identified. Through this rigorous process, the inputs of science, technology and design are well considered.

However, it is important to note that while the engineers and technocrats work in the backend of the system, the designers contribute to the front end of any system. As described by one of the greatest designers of all times, Charles Eames, 'One could describe Design as a plan of arranging elements to accomplish a particular purpose'. Instead of articulating design as an expression of art, it would be rather judicious to express design for a purpose. This can be understood from the fact that while we simply draw a line on a sheet of paper, it may qualify as artwork. However, if it is articulated with some purpose like some measurements or drawn to support any scientific principle/temperament, then it qualifies to design. While arts and aesthetics

DOI: 10.1201/9781003414957-1

1

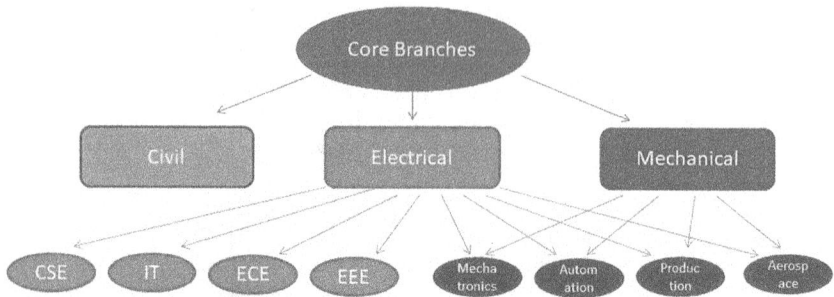

FIGURE 1.1   The core branches of engineering and their latest evolution.

are extremely important for a design, it cannot be truly termed as design until it is supported for some application-oriented work. Thus, any product, interface or tool remains incomplete without the intervention of a designer. While engineering and technology pace the growth of any product or tool, design makes it more appealing to cut into the masses.

This book takes a detailed journey from the design process to the technology inputs of daily care products for the elderly specifically emphasising re-design issues of the walking stick and umbrella taking Indian anthropometry into consideration. The approach to design is taken from the principles of design thinking.

1.1   UNIVERSAL PRINCIPLES OF DESIGN

In this section, some of the key principles of design elements that are applicable in almost all aspects are articulated. Among the first set, we need to identify that a universal design provides comfort, accessibility, adaptability and flexibility which helps improve social sustainability [1, 2]. Adaptability refers to user-friendliness in using the design or the product. Any new design needs to be tested for adaptability by the users. Further, flexibility in design aids to improve the design while taking feedback from the users or for future extension of the design to adopt newer technologies. It generally adds to a buffer quotient so that any new change in technology can also be easily accommodated in the design.

The universal design is also known as *Barrier Free design*. For improving the accessibility of any design, four parameters are crucial, namely, *perceptibility*, which deals with how information is presented without adding any cognitive load; *operability*, which emphasises that the design is operable by all; *simplicity*, to ensure that the design is not intricate for anyone to understand and *forgiveness*, so that in case the user makes any mistake in deploying the design, corrective measures should be there to re-use it. Cognitive Load refers to the amount of data or information the human mind can retain at any point of time. *It is quite well proven that any product we design for the elderly should not add to the cognitive load of elderly users. The higher the cognitive load offered by a design, the slower the speed of taking decision or the reliability of the human–machine system interaction.* The constraint of reliability of a human–machine interactive system can be explained by a simple example. Suppose the efficiency or reliability of executing any task by the user is 0.8, the efficiency or reliability of the operability of a machine is 0.8, then the total efficiency or reliability of the human–machine system interaction is $0.8 \times 0.8 = 0.64$. This means that the total efficiency of a human–machine environment is less than the individual efficiency of the human and the machine, respectively. Thus, while incorporating any design, the concept of reliability needs to be taken care of.

The Cognitive Load Theory is based on the Human Information processing model of [3]. It describes the process as having three main parts: sensory memory, working memory and long-term memory, as shown in Figure 1.2. The human mind receives input signals from all its sensory organs, whether it is through touch, feel, vision, taste, sound, etc. However, not all the inputs are relevant enough. For example, while meditating, the concentration of the person remains focused and all the adjacent sounds, noise or any other distractions are completely ignored by the human mind. Only the focus of meditation is retained and then passed to the next memory section, that is, the Working Memory. This part of the memory again either retains or discards the information received from the previous section depending on the needs. Working memory is capable of processing and holding five to nine items at any given instant of time. When the brain processes information, it categorises that information and moves it to long-term memory, where it is stored in knowledge structures known as *Schemas*. These organise information according to how one uses it.

In addition to it, there is also a *Behavioural Schema*. The behavioural schema lets one interact with the environment. The user studies generally operate to understand this schema so as to establish the human–machine or human–interaction. In fact, cognitive load is the amount of information that can be held in the working memory at any point of time. The Cognitive Load Theory also shows us that working memory can be extended in two ways. First, the mind processes visual and auditory information separately. Auditory items in working memory do not compete with visual items in the same way that two visual items, for example, a picture and some text, compete with one another [3].

This is known as the *Modality Effect*. So, for example, explanatory information has less impact on working memory if it is narrated, rather than added to an already complex diagram.

Second, working memory treats an established schema as a single item, and a highly practised automated schema barely counts at all. So, learning activities that draw upon your existing knowledge expand the capacity of the working memory [3].

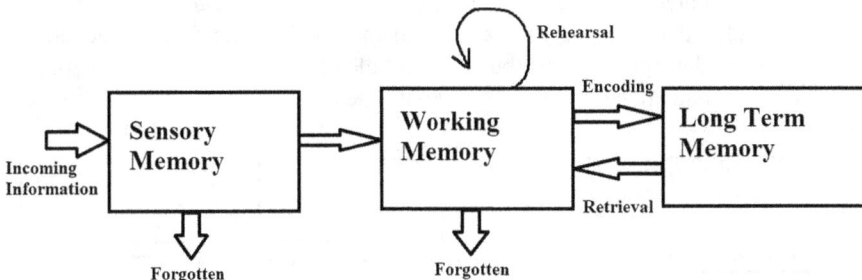

FIGURE 1.2 Information processing model based on cognitive load theory.

It should always be kept in mind that as the flexibility of any design increases, its usability decreases. This is explained as *Flexibility-Usability Trade-off*. Some other universal principles are mentioned below:

(a) For information to be presented or organised or framed, it is always advisable to adopt the *Advanced Organiser* technique. The advanced organiser is classified into two parts: First is *Expository*, where no knowledge of the design is known. This can be in the case of launch of computers against the traditional hard copy maintenance of records, teaching a new course to students who have no background or pre-requisite knowledge, etc. Second is *Comparative*, where some previous knowledge is known. For example, some advanced-level courses have prerequisites as the advanced course is based on the concepts taught previously, or any new version of a mobile handset launched by any manufacturer is based on comparative know-how than the previous. Similarly, *Chunking* of data is used to collect data sets or information of similar interests. This is prevalently used after the interview process or when a questionnaire is completed using the direct observation and analysis strategy. Collecting information or elements with the same goal is also referred to as the design principle of *Common fate*.

There are two main methods of User Study: *Direct Observation and Analysis* [4] and *Indirect Observation and Analysis* [5].

The steps in the Direct Observation and Analysis method are shown in Figure 1.3. The first step is Study Development. Study Development refers to understanding the research questions that need to be asked to frame the hypothesis. Based on it, the research questions are framed and the various data collection tools are organised. One of the many available data collection tools is selected to roll out the pilot study for the research observations. The various data collection tools are as follows:

(i) *Descriptive fieldnotes* – This is a common practice as employed in Anthropology. This technique is used where not much information is known before. It marks the complete course of an event under observation in detail. Thus, the data is filled with every smallest observation possible so that no room for leaving any information is left out. Key information such as date, time, location, people present and who recorded the information is useful for later analysis. Analysis is done at a later stage once detailed observations are made. Analytic notes to research and analyse the event may also accompany the descriptive notes, but they are clearly marked for interpretation purposes and are not to be confused with the detailed observations.

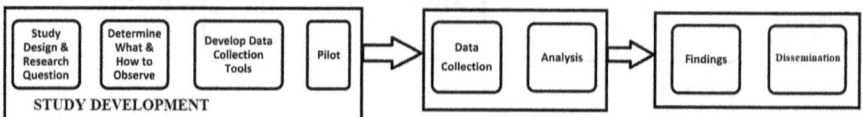

FIGURE 1.3   Direct observation and analysis process.

(ii) *Semi-structured templates* – This template comprises both open-ended and structured fields. It includes the same key information described above (i.e., date and time), then provides prompts for apriori concepts underlying the research questions, often derived from a theoretical model. These literature-based, theoretical concepts should be clearly defined and operationalised.

(iii) *Structured templates* – A structured template is like a record sheet or checklist form where what is to be observed is known beforehand. This is prevalently used in design or experiments based on the background, either psychology or Engineering. Structured observations are more deductive and based on theoretical models or literature-based concepts. The template prompts the observer to record whether a phenomenon occurred, its frequency and sometimes its duration or quality.

As referred from [4], all templates should include key elements like the date, time and observer. Descriptive fieldnotes and semi-structured templates should be briefly filled out during the observation, and then written more thoroughly immediately afterwards. Setting aside time during data collection, such as a few hours at the end of each day, facilitates the completion of this step. Recording information immediately, rather than weeks or months later, enhances data quality by minimizing recall bias. If written much later, the recorder might fill in holes in their memory with inaccurate information. Further, small details, written while memories are fresh, may seem unremarkable but later provide critical insights. For the semi-structured and structured templates, which contain prepopulated fields, there should be an accompanying 'codebook or scratchpad or scribble book' of definitions describing the parameters for each field.

The next step includes data collection. It is suggestive that more than one person should go for data collection. This helps divide the load and balances the psychometric inputs by the users. Further, multiple observers complement each other's perspectives and can provide diverse analytic insights. De-briefing and discussions before and after data collection help improve the reliability of the data collected. After the data collection, data is analysed. It is always beneficial to consider the data analysis method beforehand so as to predict a confirmatory outcome of the research carried out. The analytic plan will be informed by both the principles of the epistemological tradition from which the overall study design is drawn and the research questions. Studies using observation are premised on a range of epistemological traditions. Analytical approaches, standards and terminology differ between anthropologically informed qualitative observations recorded using descriptive field notes versus structured, quantitative checklists premised on psychological or systems engineering principles [4].

The final step of Direct Observation and Analysis is Dissemination. Dissemination means transferring the information to the required stakeholders in the format as deemed necessary by the users. Dissemination of the research study always leaves room for open questions on how the conducted research can be further extended or modified depending on the case. Whether it is a report format or journal paper or conference proceeding, the dissemination process conveys all the information which is relevant for the users of the research and camouflages the non-vital data of the research.

In the *Indirect Observation Process*, the recorded data or literature already available is analysed. The recorded data can be in the form of audio/visual clips or some textual data, facts and figures, etc. It may also involve some conversations or group discussions. The data so received is subjected to both qualitative and quantitative analysis formats. Systematic or Direct observations yield the spontaneous behaviour of the user in the study. It is interesting to note that the observations equation of Mucchielli is given as follows [5]:

$$O = P + I + Pk - B \tag{1.1}$$

Where O = Observation, P = Perception, I = Interpretation, Pk = Previous Knowledge, B = Bias. Observation thus is not possible unless what is being observed is perceivable. Perceptibility is a key concept when it comes to differentiating between direct and indirect observation. In fact, in Indirect Observation, it is only partially fulfilled from the above equation as well.

It is interesting to note that human communication is not just the passing of information. Rather, the physical and body gestures, that is, the non-verbal conduct also constitute a key part of the information transfer process. Indirect observation involves the analysis of textual material generated either indirectly from transcriptions of audio recordings of verbal behaviour in natural settings (e.g., conversation, group discussions) or directly from narratives (e.g., letters of complaint, tweets, forum posts) [5].

Various sources of materials used in the Indirect observation are stated below:

(i) *Recordings of verbal activity* – This should be recorded as it is and then re-heard on several different occasions to find out the correct interpretation based on the context of study. It may consist of different voices, single or multiple dialogues, different levels of conversations, etc.

(ii) *Transcripts of audio recordings of verbal behaviour* – This may be for an individual or for a focused group, as the case is intended for study. However, the individual may be identified clearly in the transcript of recording.

(iii) *Written text* – These could be some formal or informal written documents or text which can aid in understanding the context of the user. These may include letters, WhatsApp messages, email texts, etc.

(iv) *Objects of regular use* – These could be objects which users use in day-to-day life and for regular use. For example, for the elderly, objects could be their medicine box, walker, walking stick, etc. The emotional attachment to these objects and their value can be explained by the user in detail.

(v) *Graphical materials* – It is well said that a picture is worth a thousand words. A photograph can also act as a source of material which can be readily interpreted in the Indirect Observation process. A single photograph captures a moment, something static, but a gallery of photographs separated in time can capture the dynamics of an episode or successive episodes in the life of a person, or even a group or institution. These can be classified as *Primary* (where it is the only source of material available) or *Secondary* (where it may complement or augment a primary resource).

(vi) *Unobstructive objects or aggregates* – These comprise materials which are anecdotal in nature. For example, foot print or fingerprints or biometrics may be relevant in certain case studies.

The first step in Indirect Observation from the source of materials is to map the Qualitative Data sets. Qualitative data sets do not express anything mathematically but may refer to emotions like touch, feel, pleasure, etc. Qualitative methodology leaves ample room for interpretations. The data so captured through source material and qualitative interpretation is then put through the six stages (also known as text liquefying process). These comprise the actual understanding of the observation carried out so far. These are as follows:

Stage I – Specification of Study Dimensions. Dimension in Indirect observation refers to a distinguishable facet related to the research objective. Various User cases and studies may start with a one-dimensional approach, however, with gradual build-up of research data or material, multi-dimensional approach may also be adopted. For example, while designing any game (Digital, Computer or Hardware oriented), one starts with the approach to find out the ease of accessibility to a certain age group of users. Some games are always marked for teenagers or for various age groups like 0–2 years and 3–5 years. However, after accessing the gaming efficiency and repeatability or probability to achieve different results when played again and again, adds a new dimension. Thus, the criterion of dimension may change with accessibility or other issues at different levels of the development cycle.

Stage II – Establishment of segmentation criteria to divide the text into meaningful units. This process is also called *Unitizing*. This is extremely close to the process of chunking where the relevant set of data is extracted and the irrelevant data collected is to be filtered. The data set can be created on various principles like the Principle of Continuum and Segmentation. It is preferable to segment the data first and then establish a secondary criterion for other dimensions. However, during the whole process, it should be strictly kept in mind that the actual message of the data is not lost during the segmentation and filtering process.

Stage III – Building of a purpose-designed observation instrument. This stage identifies a clear observing instrument so that the unitised data of the previous step can be analysed with accuracy and low errors. Observation instruments can be built using category systems, a field format system, a combination of these systems, or rating scales. One-dimensional systems usually can be operated with a category-based system. However, multi-dimensional systems operate with multiple instruments in which rating scales become essential. Since multi-dimensional systems deal with multiple elements of data, field formats become very important. As referred from [5], the field format is built by creating a catalogue of mutually exclusive behaviours for each dimension. As it is not exhaustive, the catalogue is left open and is therefore considered to be in a permanent state of construction. While not required, a theoretical framework is recommendable for field format systems. Mutual exclusion of field format also refers to the accuracy of the segmentation carried out in the previous stage of data processing.

Stage IV – Coding of information. The data acquired through observations is mostly based on narratives. Narratives form the base of Indirect Observation and Analysis. Thus, before human communication is taken into account for the qualitative inputs, it is essential to also carefully map the observed narratives into well-sequenced data, which can be then tested or applied for quantitative analysis and treatment process. This step is complicated as the source of data; that is, the narratives can originate from different sources or scattered sets of sources and organizing them for statistical treatment so that they become more homogenous rather than heterogeneous at the narrative level becomes a cumbersome task.

As defined in [5], coding provides the analyst with a formal system to organise the data, uncovering and documenting additional links within and between concepts and experiences described in the data. This is also termed as 'ad hoc observing instrument'. Once the dimensions are properly selected, and the data correctly segmented and coded using the ad hoc instrument, the data set is ready to be converted for code matrices which contain information for pure quantitative information. There are multiple sets of software available in the market for coding tools. These include the CAQDAS platform, T-LAB, etc.

Stage V – Data quality control. It should be noted that Reliability alone cannot warrant the Validity of the data set. This lays the foundation of Data quality control. This is equally valid for both qualitative and quantitative methods. The two coefficients used for data quality control measure include (a) coefficients of agreement between two observers who separately code behaviours using the same dataset and observation instrument and (b) coefficients of agreement based on correlation. There are many other coefficients available; however, the mentioned two are popularly used among researchers.

Stage VI – Quantitative analysis of data. As claimed in [5], three important methods are popular, namely the lag sequential analysis, polar coordinate analysis and T-pattern detection. All three follow a statistical treatment process.

Whenever the data is time sequentially arranged, Lag sequential analysis can be performed. It is used extensively in psychology or during patterns of interactions during communication. If a system is in state X at time t, then it is more or less likely that the analysis finds out whether the system will be in state X or not X at time $t+1$, $t+2$, $t+3$, *etc.* time or not. The analysis assumes that the events are sequenced in time (a time series) but does not assume equal time intervals between events. The information for each of the categories is shown on a graph with the lags on the X-axis and the probability values (ranging between 0 and 1) on the Y-axis.

Polar coordinate analysis can be used to measure the spontaneous behaviour of players interacting in their natural environment from the perspective of a given behaviour, known as a focal behaviour [6]. This statistic provides a representative value for a series of independent values (adjusted residuals at different prospective or retrospective negative lags) to produce prospective and retrospective Zsum values which are taken from Lag Analysis.

The third method is T-pattern detection. It involves the use of an algorithm that calculates the temporal distances between behaviours and analyses the extent to which the critical interval remains invariant relative to the null hypothesis that each behaviour is independently and randomly distributed over time. It needs data, in the form of code matrices, for which the duration of each co-occurrence has been recorded.

In succession, *Inverted Pyramid* helps create the aggregation of information so as to finally reach a common point of interest in a scattered data set. This approach in design systematically distributes the load and then integrates it back to meet the common objective in the distributed loads. An example can be thought of dividing a workpiece into various chunks and then aggregating them so as to achieve the final goal. This is generally done in software development and design process, where different modules are first created, tested and then aggregated to build the final version of the software. At a system perspective, Inverted Pyramid facilitates load sharing. The inverted pyramid may be applied under *Top-Down Approach* or *Bottom-Up Approach*. In top-down approach, a problem or program is broken into different fragments or chunks known as modules, and each of these individual modules is designed or developed separately such that there may or may not be any mutual co-relation between them. This is a very common practice adopted in the software development lifecycle. The software is supposed to perform multiple and different tasks, which are all developed as individual modules. Once the individual modules are developed, the integration of each of these modules is done in the reverse sequence as they were fragmented. This is the bottom-up approach. As an example, once the individual modules of the software are done, the integration of each of these modules is done to derive the final packaged software.

However, the *Storytelling* format introduces the events in a design process as characters and then describes the processes in which the events forge ahead leading to a common goal or consensus. Information Design, Instructional strategies and design thinking approach for collecting data or field study use the above methods. The storytelling method is also deployed in advertising and marketing strategies. While teaching a course to the students, a teacher does not end up writing complicated or complex equations in the first class itself. The first class is mostly interaction based on where the teacher introduces the course and curriculum details to the students, what the prerequisites of the course (if any) are and what is expected to be delivered as a part of the course. In the subsequent classes, the foundations and basics related to the course are furbished and, then, the detailed discussion on the technical part of the course is dealt. This instructional strategy which is applied and accepted universally is based on the principle of Storytelling format. The typical steps followed in the Storytelling format includes Setting of Stage, Characters, Plot, and Invisibility of storyteller, Mood changes and Movement of characters. The steps are suitably included in the various design practices, as the case may be.

(b) Design should always foster *Cognitive Consonance*. This means that any design should lead to agreement of the senses of a user. In case there is a conflict of attitudes, belief or thoughts, it is called *Cognitive Dissonance*. An example of this is while experiencing some activity for the first time, be it the use of ATM card or credit card, experiencing any adventure sport for the first time, etc. mostly all necessary precautions are taken by the user. Yet, there is always a certain fear or uncertainty that prevails in the mind of the user. Similarly, in adventure sports, all harnesses are put for safety measures, yet the user is scared to take an immediate plunge into such activities. This effect has also been used in marketing design strategies.

Thus, it is extremely important that the *Cognitive Learning Theory* should be emphasised on [7]. Cognitive learning theory suggests that it is not enough to conclude that users respond because there is an assumed link between stimuli and response, which is based on some past history of reinforcement for s response. Instead, users develop an *Expectation* that they will receive some reinforcement after making a response. This leads to the understanding of various expectation effects which are discussed in detail under the Design Thinking Strategies. Two types of learning methods in which no prior reinforcement is present include *Latent Learning* and *Observational Learning*.

Latent Learning occurs without any reinforcement. It states that a new behaviour is learned but not demonstrated until some incentive is provided for displaying it. People generally develop a cognitive map of their surroundings. Interestingly, latent learning may permit a person to know the location of a toy store in some local market area that they might visit frequently, even though the person might not have an infant to buy toys or might not have ever entered the toy store.

According to psychologist Albert Bandura, a major part of human learning comes from Observational Learning, which is learning by watching the behaviour of another person, sometimes called the *model*. Because of its reliance on observation of others, which is a social phenomenon, it is also termed as 'Social Cognitive Approach to Learning'. It is a very common practice among children to sit on the driver's seat and then rotate the steering making a simultaneous buzz sound as if they are driving the car. However, this is also the first step in learning to drive a car. Through this approach, one may acquire positive or negative skills depending on what one is observing.

However, this is not applicable to super-specialised skills. For example to perform surgery on a patient, one needs to acquire super-specialised skill set. Interestingly, not all behaviour that we witness is learned or carried out. A key role to imitate a model is whether the model is rewarded for the behavioural conduct. For example, a child's friend may study hard, achieve good marks in studies and be rewarded because of the marks earned. This may act as a stimulus to observation in a child that if he or she also studies hard and earn good marks, he or she may also be rewarded just like the friend who becomes the model in this case.

Thus, it is equally important to understand the concept of *Mental Model*. Mental model means the mental representations developed from an experience. People tend to interact with system or with others in a social environment based on the perceived mental models developed. There are two types of Mental model, namely *System model* and *Interaction model*. System model refers to how system works. By system, one can refer to a social system, socio-economic system, product system or technical system. It is amazing to observe that when a gadget like a mobile phone is bought for the first time, the user knows how to turn on the mobile phone or turn it off or use certain icons, even though the user might not have used the mobile phone of the same brand or company ever before. In this case, the system model of basic operability of the mobile phone in the mental model of the user provides the basic aid to operate it, without any prior knowledge of the brand or the company from where it is bought.

Interaction model refers to the mental model of how people interact with the social set-up around them. When a child is born and then groomed in a family, he or she starts observing the people around him or her. Children start imitating their parents to learn basic manners and customs. The initial training on how to walk, talk and behave in front of guests is based on the interaction model which is developed in the child by observing his or her parents. Designers usually operate with very successful system models; however, for a successful design ideation and deployment, interaction model is equally important. This lays the foundation of human–machine interaction or human–computer interaction (HCI) or Cognitive Ergonomics. Actually, one must use the systems that are designed and employ them in laboratory testing or controlled environment testing and then perform field observation in order to develop accurate and complete interaction models.

A designer should always adhere to the *Principle of Simplicity*. It states that given a choice among many and different designs available which are all functionally equivalent, the simplest design should always be chosen. This helps to connect to the masses for which the designs are made. The other names of this principle are Ockham's razor or Occam's razor or Law of Parsimony or law of economy.

(c) *Colour* has an equal impact on designs. The number of colours should be limited to the capability to process in the human mindset. Too many colour options would add to *cognitive load*. Higher the cognitive load, more time it takes for decision making. The choice of warm colours and cool colours, saturation and hue can all improve the aesthetics of the design. Improper choice of colours can affect the function and form of the product or design proposed. Aesthetic colour combinations can be achieved by using adjacent colours on the colour wheel (also called analogous), opposing colours on the colour wheel (also called complementary), colours at the corners of a symmetrical polygon circumscribed in the colour wheel (also called triadic and quadratic), or colour combinations found in nature. Warmer colours are preferably used for foreground elements, and cooler colours are used for background elements. Light grey is a safe colour to use for grouping elements without competing with other colours. This is as referred from [1].

Colours are divided into Warm colours, which comprise Red, Orange and Yellow, and Cool colours which include Blue, Green and Purple. White, black and grey are categorised as Neutral colours.

- Warm colours are vivid and energetic. They give out feelings of warmth and summer. They are associated with emotions of anger and alertness.
  1. Red colour is found to stimulate hunger and demands attention in a powerful way.
  2. Orange represents creativity, youth and enthusiasm. This colour simply attracts attention.

3. Yellow represents happiness, hope and spontaneity.

   The above colour representations are used in general. However, they need to be carefully interpreted for various contexts. For example, while designing products for the Indian elderly, the use of solid warm colours is avoided as they might be interpreted as a religious sign. This in turn may affect the faith and belief of the users.

- Cool colours are calm and soothing. They liberate the feelings of spring and winter.
  1. Blue colour is associated with calmness, intelligence, therapy and meditation. The logos of most institutions are shades of blue as it radiates feelings of trust, dependability and professionalism.
  2. Green colour represents nature, growth and stability. It brings a sense of visual balance and radiates feelings of relaxation.
  3. Purple colour represents royalty and luxury. It is also associated with spirituality and mysticism.

     It has the energy of warm colours and the calmness of cool colours. In the Indian context, cool colours are seen associated with masculine nature or masculinity. It may or may not be associated with chauvinism.
- Neutral colours are black, white and grey.
  1. Black colour represents sophistication and power. It represents power and exclusiveness. When used in proper combination with other colours, it can radiate various other emotions.
  2. White colour is simple and minimalistic.
  3. Grey colour depicts formality, responsibility and maturity. It represents a strong character.

Different shades of colours can be interpreted with different meanings. Lighter shades can get attention immediately, whereas darker shades can seem more natural and mature. Using neon shades can give the colours a catchy effect but it seems very artificial. The ideal choice of colours is neutral colours as they are not associated with any gender. Black and muted silver are silent colours. Hence, they can easily merge into the lifestyle of the user.

Colour combinations along with the text message or textual data may emphasise highlighting any important or relevant text. This is termed as 'Highlighting'. While highlighting any data or information, it should be kept in mind that not more than 10% of the total text should be highlighted. There are three ways of highlighting namely through **BOLD text**, *Italic Text* and <u>Underlined Text</u>. A combination of the three may also be used to highlight any prominent text or to emphasise any important area in the text.

According to *Gutenberg's Diagram*, the pattern followed by the eyes in reading starts from top-left to right-bottom. This is depicted in Figure 1.4. When beginning to read any text or document, the eye gazes at the top-left area, which is termed as the 'primary optical area'. The reader then reads from left to right and then reaches the Strong Fallow area, which is located at the top-right of the document. After reading the first line or the row, the eye then traces the second row and first column of the textual page. Thus, the orientation of the axis of reading lies in the fourth quadrant, that is, around an angle of −45 degrees or 315 degrees. The last line beginning from the left section of the page comprises the weak fallow area. The right-bottom area of the page is termed as the 'terminal area', where the reading from the page is terminated or ceased. Interestingly, this principle is not applicable while reading in Urdu as Urdu unlike most of the other languages is right aligned. Thus, the mirror image of Figure 1.4 is applicable while reading in the Urdu Language.

While highlighting, inversion of the colours is another preferable format in the textual data. Sometimes blinking or flashing of data also catches the attention of the user. The concept of flashing to highlight any information is extensively used in the market area on the shops while highlighting the shop names.

(d) *Feedback* in any design process is extremely important. Feedback creates a loop to understand the *positive or negative* impact on the design. The feedback assessment can also be co-related to *confirmation*, where execution of a process is determined by the user. Feedback can also be *compared* to seek a better and holistic picture of any design. Feedback occurs whenever the output is routed back or looped back to the input as a part of the chain of cause and effect cycle of any system or event. Such systems are sometimes also called *Self-Regulation Systems*, as they have the capability to re-correct themselves on account of the loop back of the output generated along with the input provided. Positive and negative feedback are also employed in electronics engineering. Oscillators are circuit components that use positive feedback through the Operational amplifiers, whereas active filters in Operational Amplifiers use negative feedback concepts.

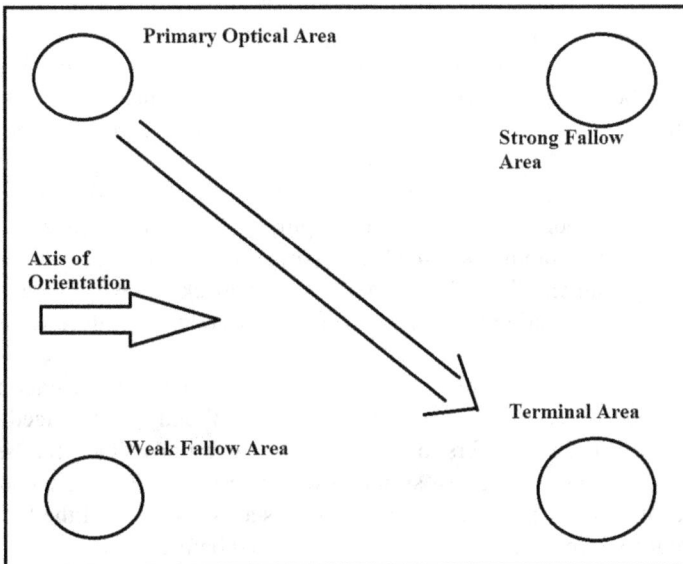

FIGURE 1.4 Gutenberg's diagram to showcase the reading pattern of the human eye.

In design, positive or negative feedback may be defined as the valence of the action or effect that alters the gap, based on whether it has a happy (positive) or unhappy (negative) emotional connotation to the recipient or observer. It is extremely important to take feedback from the user in order to improve any design input. As a classical example, in earlier versions of the operating systems, whenever the user deleted some file or folder or document, it was removed from the memory of the system or the computer. There were several occasions when a user might delete some document due to an oversight also. Thus, based on the feedback of the user, a dialog box now pops up which asks the user 'Are you sure you want to delete the file or folder?' This is especially true, whenever permanent delete option is executed in Windows. The window pops up the question, to which the user needs to make a decision in terms of Yes or No and then the file or folder is deleted or restored in its original position.

Another classical example is the electric iron or press, to iron clothes. In electric irons, a bi-metallic strip is used to ensure that as the iron becomes hot, it must be cut off to avoid any damage or fatally hurt the user. However, key information is after the cut-off, the iron starts becoming cool. Thus, whenever the user uses the iron again, the cooled press cannot be used and the user does not know when to use the iron again. This is avoided by using a temperature or cooling strip which gives feedback to the user as to when the iron can be reused again after it is cooled to cut-off again. Thus, feedback in the design set-up helps also improve the interaction of a human–machine or human–system interaction. Feedback can be modelled qualitatively or even quantitatively. Further, Likert scale mapping of the feedback is also a common approach adopted in user-centric design practices.

Another form of feedback is corrective feedback. This kind of feedback is employed in the educational or learning kind of system. The feedback in this is based on the performance evaluation of the learner entity, which can be formal or non-formal feedback. For successfully deploying corrective feedback, it should have the traits to be non-evaluative, supportive, timely and specific. This kind of feedback can be sometimes verbal or in the form of grades, marks, or letters or represented through a percentage. Typical examples where corrective feedback is employed are assessment patterns adopted in school education or performance evaluation of an employee in a company. Sometimes this is also marked as a part of a Key Performance Indicator (KPI) of the employee. Corrective feedback begins in early childhood with motherese, in which a parent or caregiver provides subtle corrections of a young child's spoken errors. Such feedback, known as a recast, often leads to the child repeating his or her utterance correctly (or with fewer errors) in imitation of the parent's model.

In recent times, two more new terms are used quite often to understand the feedback concepts better. These are 'negative feedforward' and 'positive feedforward'. The negative feedforward refers to comments or suggestions which can be used as constructive elements to improve system-level performance for future endeavours. Positive feedforward, on the other hand, works as an affirmation of the fact that the system progress is in the desired direction as decided before.

(e) Any process or design or product undergoes the four stages of *Development Cycle* namely:

(i) *Requirements* which assess the needs of the process or design or product. For assessing the requirements, a need analysis can be done or a user study may be carried out. This is the first stage and so critical reasoning based on direct and indirect observational and analysis methods may be employed.

(ii) *Design* refers to the ideations based on the requirements and needs. The design at this stage undergoes various ideations and concept building. Ideations can be on hand sketches or digital rendering. Various digital rendering tools are available on which concepts can be built.

(iii) *Development* refers to prototyping and realizing the design in physical self. There are different types of prototypes which may be useful for the experiment purpose. They are discussed in the next section.

(iv) *Testing* means taking the developed prototype in real-time environment and seeking its feedback to test efficacy of the design. The testing can be done in a controlled environment set-up like lab scale or in an uncontrolled environment like the practical workspace or the field-testing area. However, feedback, as discussed previously, always renders a helping hand to improve the overall development cycle.

It should always be noted that any design process or product developed cannot be optimally designed for all. But if it meets the requirements of the majority, then the design is said to adhere to *Satisficing principle*. Optimal solution may not be acceptable to all users. This principle is also stated as 'Best is the enemy of good principle'. This is generally applicable under three conditions, namely very complex problems, time-limited problems and problems for which anything beyond a satisfactory solution yields diminishing returns. It is interesting to note that the same principle is used in industry while rolling out a technology or a product. It is quoted that in a test of time versus technology, technology always wins. It can be understood by an example of the launch of an operating system. Whenever a new version of the operating system is to be launched, a beta version is given to users whose feedback in terms of lacunae in the technology is collected. This stage is called beta testing of software.

Once the initial beta testing results are collected, the operating system is corrected and improved and shortly it is launched in the market. Industry people do not wait for the most optimal result of the operating system to be completed. Even with small errors or pitfalls, the technology is rolled out. Slowly, minor corrections, upgradation and improvements are introduced, which adhere to uplift the technology for wider user coverage. If the company waits for the most optimal results or optimal software or operating system to be developed, a competitor of the company may roll out the newer version and capture the market, leaving no scope for the original company which is waiting for the optimal design to be completed. This process is in tandem with Satisficing principle so that maximum users can be covered leaving scope for minor upgradation, which can be done at any later stage also.

Another example is the launch of the 2G, 3G and 4G mobile services. In India, the services were launched by all competitor companies even before the optimal solution of technology was reached. Once the network reached beta testing with working condition for a network operator, the services were launched without a delay. The minor upgradation and improvement in coverage area and data rates or data speed were all being slowly jacked up with pace of time. However, since there are too many competitors in the network operators, the latest technology is launched first with satisficing principle instead of waiting for the optimum solution of the technology to be reached.

Further, increasing too many alternatives in any design increases the decision time. This is *Hick's Law*. Hick's Law is extremely crucial during emergent situations where a person or a user needs to take a decision with minimal cognitive load and high reflexes. However, if options increase in any design, the cognitive load to take a decision also increases. This is also applicable for various products for the elderly. If too many options are provided to the elderly, the cognitive load on them also increases, thus the usability of the product in context of the elderly also decreases. To perform any task, there are four stages involved namely:

(I) *Identify a problem or a goal* – In this case, the problem statement needs to be identified, for which the solutions need to be worked out. Problems and research gaps lead to the pathway to propose necessary solutions for improvement of the task.

(II) *Assess the available options to solve* – Once the problems are clearly laid out, the feasible set of solutions from the already available options is assessed. In case, the regular available solutions do not work out, then an innovative method or solution set needs to be proposed and worked upon.

(III) *Decide option* – From the multiple set of solutions available, the user then needs to make a choice to use the appropriate method for solution. It is at this stage that Hick's law is applicable. More the choices available to the user for deciding any solution, more is the cognitive load, thus leading to a delay in taking the correct decision.

(IV) *Implement option* – Once the correct decision is made, the solution or the option needs to be executed or implemented.

It is thus suggested that any design both for software and hardware can be *chunked and allowed prospect-refuge strategy/progressive disclosure*. Whenever many alternatives are to be offered, it is better to group them in chunks and then they may be regenerated as a part of some module which can provide the alternatives as the product or design is further explored by the user. This is extremely common in software designs using the options of menus, drop-down menus, scroll bar, etc. Prospect refuge or Progressive disclosure strategies are used when too many options may be provided to the user based on some need. For example, if someone wants to save a word document in some other format, the user needs to first access the File drop-down menu and then select 'Save As' option and then select the format in which the file needs to be saved, that is, in pdf format or any other format. Through this way and the various other drop-down menus available as options to the user, the concept of progressive disclosure in implemented. It ensures that the 'Save As' option is executed only when the user needs it. It is not present at all times adding to cognitive load in taking decision for saving a file.

While designing any cognitive-based system, it should be kept in mind that memory of *recognizing things is better than memory to recall things*. This is of extreme importance whenever any product is to be designed for geriatrics and self-care. Recognition memory is obtained through exposure and does not necessarily involve any memory about origin, context or relevance. It is simply memory that something (sight, sound, smell, touch) has been experienced before. Recall memory is obtained through learning, usually involving some combination of memorization, practice and application. Recognition memory is also retained for longer periods of time than recall memory. This is also referred from [1].

Von-Restorff effect or *Novelty Effect* or *Isolation Effect* is a phenomenon of memory in which noticeably different things are likely to be recalled than common things. This is explained in two different ways. The first is Differences in Context: For example, in a sequence of EZQL4PMBI, '4' is likely to be remembered since it is different. If '4' is replaced by 'T', it will not be remembered. The second is Differences in Experience: for example, people often remember their first day of college or major event in life.

A phenomenon in which mental processing is made slower and less accurate by other competing or pre-existing mental processes is termed as 'interference effect'. There are different types of interferences which are listed as follows:

(I) *Stroop interference* – It occurs when an irrelevant aspect of a stimulus triggers a mental process that interferes and convolves with processes involving a relevant aspect of the stimulus. For example, using Calculator on mobile may mislead to entering contact number on mobile. It sometimes has been observed that people tend to do calculations on the keypad of the mobile phone or try to save a mobile number in the calculator section of the mobile phone.

(II) *Garner interference* – It is an irrelevant variation of a stimulus which triggers a mental process that interferes with processes involving a relevant aspect of the stimulus. For example, searching for Scientific Calculator features in normal calculator. In a hurried situation, people/students using scientific calculators tend to use trigonometric or other advanced-level calculations in a simple calculator.

(III) *Proactive interference* – In this type of interference strategy, existing memories interfere with learning methods or processes. For example, in learning a new language, people apply grammar of native language. Since the native language or mother tongue is the first language of any person, it is observed that while learning a second or third language, people tend to apply the grammatical rules of the native language while learning the newer one.

(IV) *Retroactive interference* – In this case, learning interferes with existing memories. For example, learning new phone number can interfere with phone numbers already in memory.

All these possible interference effects affect the cognition, recall or recognition capabilities of any user. This is in turn has to be strictly taken care of during any design process be it products, instructional strategies, interaction, visual design, etc. This is referred from [1].

1.2 WHAT IS DESIGN THINKING?

Design thinking is a human-centred approach to innovation – anchored in understanding customer's needs, rapid prototyping and generating creative ideas – that will transform the way one develops products, services, processes and organizations. By using design thinking approach, one can make decisions based on what users want instead of relying only on previous data or non-scientific principles based on instinct instead of evidence. This is referred from [8, 9].

The three main pillars of design thinking approach are shown in Figure 1.5. *Desirability* assesses the needs of the users. The requirements are all mapped in this quadrant. This can be executed through direct observation and analysis techniques like interviews and questionnaires. *Feasibility* means that the problems can be sorted out or not. And what is the satisfying outcome as compared to the optimal output? Feasibility study also lays down that if there is no direct feasible or possible solution, then how is the problem to be evaluated with other alternative solution modes? *Viability* studies that feasible solution, as proposed in the last step, is how long likely to survive. Whether the viable solution will be acceptable or not by the users is also tested alongside.

As quoted by Sandy Speicher, IDEO CEO, 'Design thinking is not limited to a process. It's an endlessly expanding investigation'. Generating concepts and

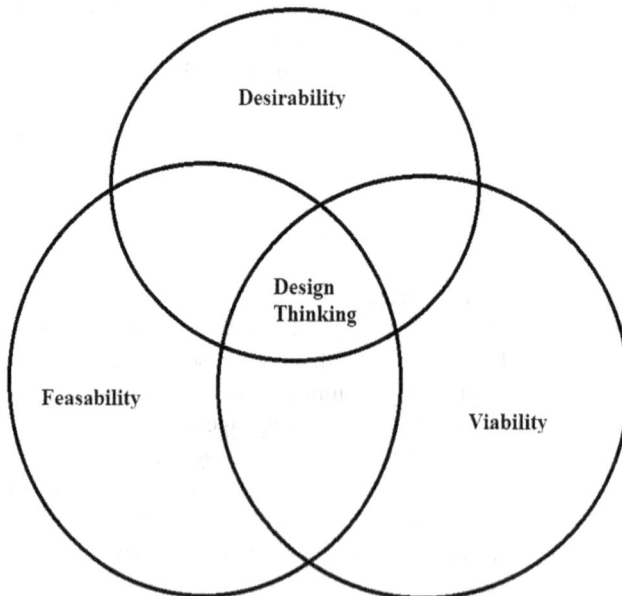

FIGURE 1.5   The three main principles of design thinking.

ideations for satisficing solution is key to achieving design thinking process. However, all three pillars need to be addressed in the process of concept generation and ideation. Design thinking process is *User-driven activity* where user inputs to find possible solution is taken into account in every stage of development cycle and also product lifecycle. The concepts of design thinking are not applicable to design stages or to honing management skills. The key focus of this approach is vested in the user only.

Interestingly a pertinent question is asked why the approach is not engineering thinking or technology thinking. It is important to understand that while engineering and technology refer to some protocols or affirmative steps that need to be followed by the user, design thinking takes the inputs from the user so that the process or design or the tool under investigation becomes friendlier to use. As referred from [10–13], there are four principles of design thinking, namely:

(I) Human rule: It is important to observe that irrespective of the context in which any design is to be evaluated, the design activities are considered social in nature. This in turn brings the focus of the entire design procedure towards the human.

(II) Ambiguity rule: Ambiguity means any possibility to understand something, some concept in more than one way. This may lead to some confusion and errors. Ambiguity stands inevitable as the subject of interest in Design Thinking approach is the user, and it cannot be removed or oversimplified. Thus, to simplify this problem, experimenting at the limits of one's knowledge and capability is important in being able to visualise things differently.

(III) Re-design rule: Interestingly, all designs are basically re-design. While, with passage of time, technology and socio-economic circumstances may change and evolve, the basic human instincts or needs remain the same. Thus, design evolution only means re-design of the concept so as to cater to better needs and requirements.

(IV) Tangibility rule: Until and unless the design leads to a tangible output, the re-design or the design remains of no use. Tangibility refers to a valid and solid outcome which can lead to improvement in the overall quality of the solution provided towards the problem under the context of consideration.

Though design thinking places the user as a priority, it is much different from user-centric design. While the common attributes between both are empathy, problem solving, iteration and collaboration, there is still a major difference in the context of approaches of the two methods. The user-centric design associates itself with a deep empathy connected to the user. While creating solutions, user inputs and their needs are sought as feedback at all stages of the design iterations. This approach is most desirable when a much-demanded product or technology or design intervention is needed for a specific set of users or a categorical set of users. For example, user-centric design is of key focus for medical equipment designs or health care products for geriatrics or elderly, etc. These are specialised designs, however, much needed for assessment of the patients by the doctors. Thus, a deeper study using user-centric design becomes extremely vital.

Conversely, design thinking approach is rendered at places where user interests also take technological feasibility and business goals into consideration. Design thinking deploys different reasoning capabilities to identify and solve complex problems that may affect product design or organizational policies, processes and functions. As an example, improving the design of a laptop or mobile phone or smartwatch or any useful electronic gadgetryall fall in the paradigm of design thinking approach. These products are mostly applicable to a generalised set of users rather than any customised set of users. Based on the approach, there are different steps or methods or approaches to implement the design thinking strategy. These are all discussed in subsequent sections.

1.3   DESIGN THINKING METHODS

The five-stage Design Thinking model proposed by the Hasso-Plattner Institute of Design at Stanford (d.school) is used as a benchmark example. Empathise, Define (the problem), Ideate, Prototype and Test [12]. This is explained in detail as follows:

(A) *Empathise* – Empathise means to understand and share the feelings of others as if we have undergone the same circumstances. Otherwise, it is Sympathy. Design Thinking uses the concept of Empathise rather than Sympathise. This allows the designer not to assume anything of their own and keep the design solution User-friendly. This is also implemented using *Empathy Maps* discussed later in this section.

(B) *Define (the Problem)* – In this stage, the data gathered in the Empathise stage is analysed to understand the clear problem definition. The define stage articulates the user inputs and what problems need to be addressed.

(C) *Ideate* – In this phase, the designers create ideations or generate various ideas which can be the possible set of solutions to the problem. To 'think outside the box' in order to identify new solutions becomes the key constraint in this phase.

(D) *Prototype* – At this point, low-cost, scaled-down or scaled-up, or one-is-to-one prototype is developed which is then used for testing phase in a controlled environment. The suggestions and feedback received in this stage are then again incorporated until the satisficing prototype is formed. The controlled environment means lab scale, within the organization, etc. Prototype means the use of simplified and incomplete models of a design to explore ideas, elaborate requirements, refine specifications and test functionality. There are three types of prototypes, namely Concept, Throwaway or Evolutionary. Concept prototype is meant only to be developed as a model for the purpose of understanding. This kind of prototype can be made from leftover or simple materials which may not be original materials out of which the final product is to be made. This can be made with materials like wood or thermocol and may be developed further as the design research progresses. Throwaway prototypes are meant to be used one or two times and then not to be used at all. These prototypes can be made from any material where only minor whereabouts are to be checked before the final prototyping is done. Evolutionary is meant for the comprehensive understanding of the product and can be taken for field testing under both controlled and open environment. The feedback taken from this prototype is mostly minor corrections, which can be incubated in the final product to be launched.

(E) *Test* – At this stage, the final testing is done and based on feedback received. The developed product may be summoned back to re-design or re-define stage also. The Testing stage results the success rate of the design and the products so envisaged.

The design thinking model is implemented using various methods. Some of the important methods are discussed in this section. Among the most powerful tool is the *Diamond Map Method*. This is as referred from [12]. The design process is divided into four distinct phases – Discover (Research), Define (Insights), Develop (Ideate) and Prototype – the Double Diamond is a simple visual map of the design process. Throughout the process, a number of possible ideas are created (divergence) before refining and narrowing down to the best idea (convergence). This is clearly shown in Figure 1.6.

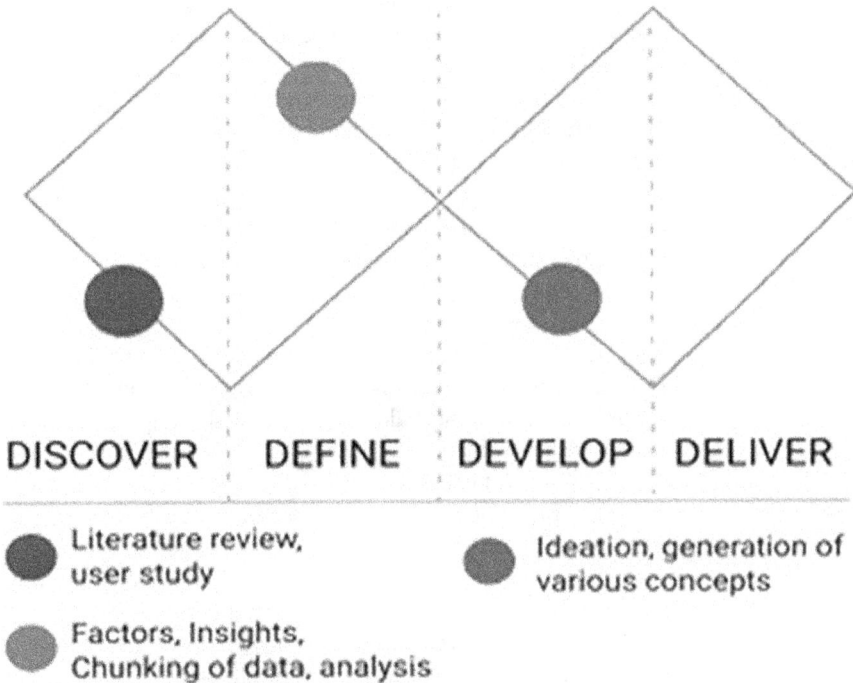

DISCOVER   DEFINE   DEVELOP   DELIVER

Literature review, user study

Ideation, generation of various concepts

Factors, Insights, Chunking of data, analysis

FIGURE 1.6 The double diamond map with all the stages depicted clearly.

The various elements of the Double Diamond model are discussed below:

1. *Discover (Research):* The design process according to the double diamond model begins with the discovery of user needs, current design and choice of various methods for the same. This is depicted in Figure 1.6. Based on the available data, the various methodologies followed are given as under:

    Shadowing: Shadowing is the behavioural observation of a user in their natural environment that provides direction for further user research. Shadowing as a tool is much stronger than interviews and questionnaires as it refrains from bias in answers under the Hawthorne effect. The psychological effects, including Hawthorne effect, are explained in the later section.

    Interviews: Semi-structured interviews can be conducted with all the participants of a study to collect qualitative and quantitative data among different categories: objective measures of the environment, residents' perceptions and attitudes about the environment and pain points along with habit formation. In a lot of cases, the interview can be a better alternative to the questionnaire or a narration.

    Questionnaires: A questionnaire is a set of questions for obtaining statistically useful or personal information from individuals. It has both qualitative data and quantitative data for collection of appropriate results.

    Qualitative data: Qualitative data helps understand the various requirements and aspirations of the users and allows insights from behaviour. This data involves non-numerical data like text, video or audio. It can be used to gather in-depth insights into a problem or generate new ideas for research.

    Quantitative data: Quantitative data helps understand the patterns of usage, pain or discomfort and activity. This data consists of numerical data. It can be used to find patterns and averages, make predictions, test causal relationships and generalise results to wider populations.

    Likert scale: Qualitative data describing the discomfort felt or any emotional relationship/feeling can also be quantified using the Likert scale. It is a psychometric scale commonly involved in research that employs questionnaires. It is the most widely used approach to scaling responses in survey research. The minimised Likert scale is between 0 and 2. The scale can be of 7- and 10-point scale also. It is equally important that while the data is collected, error should be avoided. For this purpose, the following steps can be taken:

        Mom's test: This test suggests never clouding the interviewee's perspective by the judgements or suggestions of the interviewer. One should stay neutral in tone and conversational while questioning so as to avoid control or influence over the thoughts of the interviewee.

        Rechecking: The user is asked to repeat their answer later on or asked the same questions again in a different order to confirm that what they have said was in fact how they feel. Only when both are in tandem, the recorded observations can be held valid.

        Testing for bias: Users are interviewed individually; they are asked a lot of 'whys' and 'hows' to really know if their answers are logically supported or a mere whim.

2. *Define (Insights):* This refers to defining the areas of interest and development by analysis of data collected in the previous stage. The problem statement is now chunked, and all factors are analysed in depth. Chunking, top-down approach or other techniques are used to understand the responses made by the users. The detailed analysis of the results and the user study conducted lead to the stage of ideation.

3. *Develop:* The develop stage is dedicated to ideation and generation of concepts. In this stage based on previous data collected, multiple concepts and renderings are done to achieve possible solution to the problem statement in designs. The ideation can be done on simple pencil sketches or even on digital media or through software tools like CATIA for three-dimensional rendering.

   The develop stage may also lead to multiple low-cost prototypes to be developed for further assessment and user study in a controlled environment. The prototypes may be of concept of throwaway stages. Sometimes evolutionary prototypes are also proposed by this stage depending on the ideation proposed.

4. *Deliver:* The last step of double diamond map helps achieve the product or the concept which satisfies the user requirements. A detailed user study or field study is carried out in order to test the efficacy of the products so developed. The outcome can be an evolutionary prototype or a product itself.

Another important tool of Design Thinking approach is the *Empathy map.* An empathy map is a collaborative visualization used to articulate what we know about a particular type of user. The word 'empathy' means that one person can connect to the other based on a similar experience undergone. This is thus considered a step closer to connect to the users instead of extending sympathy to the users. Hence, the name was possibly coined as 'Empathy maps'. The key focus in this tool is to empathise with the user at all levels of understanding.

It externalises knowledge about users in order to:

1. create a shared understanding of user needs
2. aid in decision making

Traditional empathy maps are split into four quadrants (Says, Thinks, Does and Feels), with the user or persona in the middle. Empathy maps provide a glance into who a user is as a whole and are not chronological or sequential. This is depicted in Figure 1.7. After the chunking of the data is done, the information is further placed in the empathy map to understand the main requirements. Empathy Map can be executed for a single user or multiple users as well. This is explained in detail along with the case studies of Walking Stick and the Umbrella re-designs in the subsequent chapters.

The design thinking process also has a very important tool for analysis known as the *Journey Map* [14–18]. A journey map is a visualization of the process that a person goes through in order to accomplish a goal. The journey map is an effective tool to compile the series of user actions into a timeline. Then, the timeline is fleshed out with user thoughts and emotions in order to create a narrative. This narrative is condensed and polished, ultimately leading to visualization.

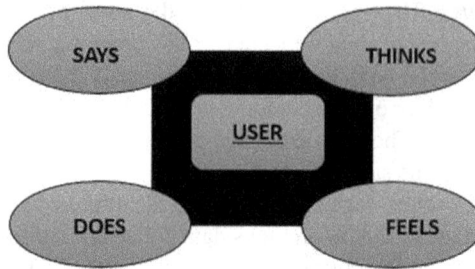

FIGURE 1.7 An empathy map representation of the design thinking approach.

Most journey maps follow a similar format: at the top, a specific user or a specific scenario, the corresponding expectations or goals in the middle and high-level phases that are comprised of user actions, thoughts and emotions, and at the bottom, the takeaways: opportunities, insights and internal ownership. The key elements of a journey map are as follows:

1. Actor or the persona, on whom the complete journey map is based. The actor/actors are also the users of a product, users using a User Interface (UI) Design or User Experience Design (UX). The design is meant to satisfy the dire needs of the actors or the users.
2. Scenarios and expectations describe the situation that the journey map addresses and is associated with an actor's goal or need and specific expectations. Scenarios can be real (for existing products and services) or anticipated – for products that are yet in the design stage or yet to evolve.
3. Journey phases provide organization for the rest of the information in the journey map (actions, thoughts and emotions). This is step-by-step movement from one task to the other in order to accomplish a clearly laid out goal and objective.
4. Actions, mindsets and emotions that the actor has throughout the journey and that are mapped within each of the journey phases. Actions are the actual behaviours and steps taken by users. This component is not meant to be a granular step-by-step log of every discrete interaction. Rather, it is a narrative of the steps the actor takes during that phase.

   Mindsets correspond to users' thoughts, questions, motivations and information needs at different stages in the journey. Ideally, these are customers' exact words spoken or written from research. Emotions are plotted as a single line across the journey phases, literally signalling the emotional 'ups' and 'downs' of the experience.
5. Opportunities (along with additional contexts such as ownership and metrics) are insights gained from mapping. These are necessitated in order to provide the most optimised solution to the user.

The major advantage of a journey map includes a clear representation of the user expectations along with a mental model to fit all the requirements of the user. Also, the shared artefact resulting from the mapping can be used to communicate an understanding of your user or service to all involved. Journey maps are effective mechanisms for conveying information in a way that is memorable, concise and that creates a shared vision.

Figure 1.8 shows a typical example of how journey maps are plotted. This is also deliberated in the technological intervention of the products using process flow to integrate technology into the products designed for elderly in the book. The same is deliberated in the subsequent chapters.

The design thinking tools also rely on the *Expectation effect*. The expectation effect means that perception or behaviour can change depending on expectations from self or from others also. There are five major expectation effects:

1. *Halo Effect* – the tendency for an impression created in one area to influence opinion in another area. For example, in a recent trend, opening of a movie is considered a success rate. This was not the case in earlier days a when movie's success was determined by how many weeks it sustained in the theatre. Whether the movie hits a silver jubilee (survival in the movie theatre for 25 weeks) or golden jubilee (survival in the movie theatre for 50 weeks) etc. decided the success of the movie. However, as per the current trends, opening day collection gives an impression to the spectators of whether the movie is worth watching or not.
2. *Hawthorne Effect* – the alteration of behaviour by the subjects of a study due to their awareness of being observed. This is of extreme importance while referring to interviews. In an interview, if the user is made conscious, then

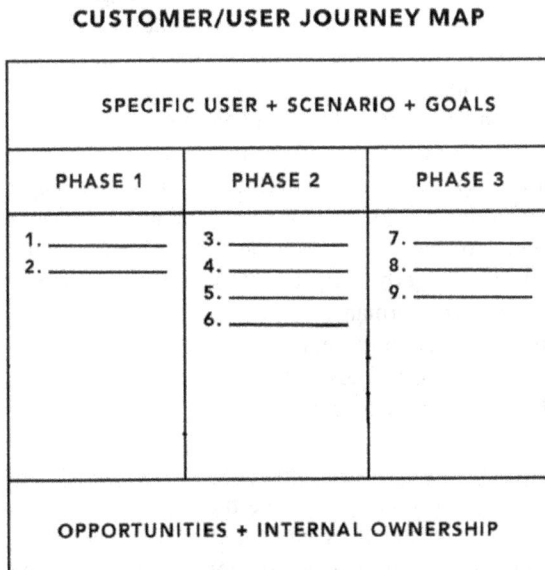

CUSTOMER/USER JOURNEY MAP

SPECIFIC USER + SCENARIO + GOALS		
PHASE 1	PHASE 2	PHASE 3
1. ____ 2. ____	3. ____ 4. ____ 5. ____ 6. ____	7. ____ 8. ____ 9. ____
OPPORTUNITIES + INTERNAL OWNERSHIP		

FIGURE 1.8   Journey map as design thinking Tool.

the honest response of the user gets sabotaged to some extent. For example, while taking interviews on some financial matters, we should never start by knowing the financial capability of the interviewee. We should ask some general questions first and then towards the end, the question on someone's financial standing can be asked.

3. *Pygmalion/Rosenthal Effect* – The phenomenon whereby higher expectations lead to an increase in performance. It is worth interest that some courses in certain fields of studies are always presumed to be difficult by the students. For example, in Electronics and Communication engineering, Electromagnetics is considered as a tough subject. If, however, the teacher from day one encourages the students to perform well, the likelihood that the results of the subject are good shall be expected to be comparatively high.

4. *Placebo Effect* – The psychological phenomenon, in which the recipient perceives an improvement in condition due to personal expectations, rather than the treatment itself. This effect has been the epitome of strength during the Covid-19 pandemic. At a time, when vaccines were not available, it was encouragement by peers, doctors and relatives that was most needed by patients to combat the disease.

1.4 ANTHROPOMETRY AND ERGONOMICS

As is well defined in [19], Ergonomics is the relationship between human, product and the environment in which human exists. The environment can be controlled or uncontrolled, and indoor or outdoor; it can also be field study. However, the focus of ergonomics remains on human and system interaction. Ergonomics is broadly categorised into three sections, namely *Physical Ergonomics, Cognitive Ergonomics* and *Organizational Ergonomics.* Physical Ergonomics refers to physical measurements and any feature which can be measured directly. Cognitive ergonomics refers to mind mapping and how cognition contributes in human–system interaction. Organizational ergonomics considers the structures, policies and processes of any organization. The goal of organizational ergonomics is to achieve a harmonised system, taking into consideration the consequences of technology on human relationships, processes and organizations at a macro level. Organizational Ergonomics is sometimes also called Macro ergonomics. There are five aspects of ergonomics namely, safety, comfort, ease of use, productivity/performance and aesthetics. Prominent ergonomic principles include the use of neutral posture, rotating tasks to avoid overwork of muscles/other tissues, use of proper handholds and proper gripping technique and proper lifting/carrying/pushing/pulling procedures.

Among the physical ergonomics, Posture analysis using REBA (Rapid Entire Body Assessment) or RULA (Rapid Upper Limb Assessment) scale is of significance. The REBA is a tool used to evaluate the risk of musculoskeletal disorders (MSDs) associated with specific tasks within a job. It is a whole-body screening tool that follows a systematic procedure to assess biomechanical and postural loading on the body. Similarly, the RULA is ergonomics-based workplace risk assessment tool that allows you to calculate the risk of musculoskeletal loading within the upper limbs and neck. RULA is easy and quick to use and does require expensive equipment to complete. Thus, ergonomics provides a comprehensive scientific study using statistical measures to quantify and understand the behaviour of human–system interaction.

In the process of human interaction with the system, the anthropometry details become extremely crucial. This is in fact an integral part of the study of Physical Ergonomics. Anthropometry means the scientific study of the measurements and proportions of the human body [20]. Specifically, anthropomorphic measurements involve the size (e.g., height, weight, surface area and volume), structure (e.g., sitting vs. standing height, shoulder and hip width, arm/leg length and neck circumference) and composition (e.g., percentage of body fat, water content and lean body mass) of humans. Anthropometric measurements in the field of ergonomics are obtained in a variety of positions, including sitting, standing and lying down, as well as various derivatives of these poses (e.g., arms stretched out, hands on a table and arms raised above the head and grip). The core elements of anthropometry are height, weight, head circumference, body mass index (BMI), body circumferences to assess for adiposity (waist, hip and limbs) and skinfold thickness. There are two types of Anthropometric measurements: Static and Dynamic.

Static or Structural Anthropometry includes Skeletal dimensions, which measure distance of bones between joint centres and include some soft tissue measures in contour dimensions (include the wobbly stuff that covers our bodies – muscle, fat, skin, bulk). This does not include clothing or packages or packaging. Dynamic or Functional Anthropometry refers to distances measured when the body is in motion or engaged in a physical activity. This includes reach (e.g., could be arm plus extended torso), clearance (e.g., two people through a doorway) and volumetric data (kinetosphere). Any human body is characterised based on the amount of bony material, fats or muscles. This classification of human body is called *Somatotyping*. There are three main types of body Somatotypes. An ectomorph is a very skinny, lean and thin person. An endomorph is a bulky person with a lot of fat and fat deposits in the human body. A mesomorph is a muscular person.

A key factor representing the Anthropometric data is by using percentiles. Percentile is discussed in the section of Statistical tools and treatment on data in the subsequent chapters.

Ergonomic designs are tested with a series of experiments which involves:

(A) Obtaining anthropometric measurements to derive 'ergonomic dimensions' of posture and movement
(B) Recording the subjective feelings of comfort that the individual experiences when using the equipment
(C) Evaluating the ability of the instrument to perform the desired activity

This is as referred from [21]. Detailed data on Indian anthropometry is compiled in [20]. This book covers the anthropometry details of Indian people covered from 23 cities in India within the age group 18 to 60 years.

1.5   DESIGN AS APPLICABLE TO PRODUCTS

In this section, some of the Design principles which are much relevant to product design shall be discussed. The *Product Lifecycle* [1] comprises the following four stages:

(i) *Introduction* – A stage which is defined as the birth of the product or when the product is launched. This might sometimes overlap with the *testing phase* in the design cycle. The design inputs should be to ensure that the product is well received among the customers or end users.

(ii) *Growth* – This stage ensures that the support system necessary for end users is maintained by the design team. There is also a strong likelihood that a glitch in this phase of the product may result in the collapse of the product itself. Thus, beta-testing of the prototypes plays a pivotal role in the growth phase of the product lifecycle.

(iii) *Maturity* – At this stage, the product reaches its peak and receives stiff competition from the competitors in the market. Further which there is a fall in the sales observed.

(iv) *Decline* – During this stage, the product sales come down rapidly and sometimes the product is driven out of the competition league completely. The design should now focus on least maintenance cost of the products.

It is interesting to note that a high percentage of effects or repercussions in any system are caused by the low number of variables. This is called as *80/20 rule*. This is also true in case of product design. One also needs to pay attention to the *Aesthetics* of the product. It is observed that high-aesthetic products are perceived to be easier than low-aesthetic products. The function of the two products from different companies may be the same, however, the user chooses a better aesthetic product for comfort. This is also termed 'aesthetic-usability effect'.

There is a popular term in Product design known as *Form Follows Function*. For example, while designing anything related to grip, we choose a cylindrical shape for a grip. This can be handles of scooters, walking sticks, walker handles, doorknobs, etc. Thus, the form over here is the cylinder and the function is related to grip. This may also be related to *Affordance*, where the physical characteristics of a product, influence the function of the product.

An important principle for products is *Constancy*. This means that the objects are perceived to be unchanging despite changes in sensory inputs like vision, loudness, etc. This is classified as *size constancy, brightness constancy, shape constancy and loudness constancy*. Even if distance changes, objects might appear of the same shape or brightness or shape, etc. A walking stick seen from far appears to be of almost similar size when viewed closely.

Similarly, *Constraints* of a product limit its functioning. Constraint also refers to limits of the actions or performance of any system. This is divided into two segments, namely *Physical constraints and Psychological Constraints*. Physical Constraints limit or redirect actions through path, axes or barriers. An example to understand it is for Path planning of robots, where all three axes using the sensor inputs of the robot decide the motion and path of the robot. Symbols, Conventions and Mapping limit the Psychological constraints. The best example is traffic lights, where red colour means Stop, Yellow means Ready to go and Green means Go or move in any traffic signal.

Any product must be *Cost effective*. The cost-benefit factor is an assessment that whether it is truly worth buying a product or discarding it. Another important tool is *Mimicry*, which is an art of copying properties of familiar environments/objects and then affording them in the design to perform certain tasks. Mimicry can be of *Surface* (making a design look like something else), *Behavioural* (making a design act like something else) or *Functional* (making a design work like something else). The surface mimicry is deployed in computer icons. The folder icon on the desktop resembles the actual two-dimensional image of a folder. Behavioural Mimicry is deployed in toys. In order to draw attention and make it attractive for kids and infants, some toys have touch sensors, which, when fiddled or touched, generate the voice of that animal. Functional mimicry is observed in touchpad phones. Mimicking the keypad of an adding machine in the design of a touch-tone telephone is one such example. The above examples are just for the sake of knowledge to understand the enormous number of applications of mimicry.

Products can also be *modular* in nature. Dividing the product into segments and sub-segments helps improve the overall functionality of the product. In fact, multifunction products can be divided into various modules and each module when accessed can further lead to newer functions. This is also referred to as *Prospect-Refuge* in design. This means that areas which need not be accessed; for example, complex circuits in mobile phones for users can be totally covered known as refuge, whereas the other functions, like accessing various apps/software may be modular and can be used only when accessed. This is called Prospect.

Every product generally has a *Defensible Space*. These are territorial markers of ownership in a product. This is classified as *Territorial* (in a house, we have space for parking, garden, etc.), *Surveillance* (in order to protect/handle an electronic device, sensors may be put up in the outer periphery of the product to seek inputs on its handling and repercussions of handling, thus generating alarm, etc.) or *Symbolic* (which may not be visible but is perceived as a defensible territory).

Redundancy literally means a situation where something or some part or some area or some text is unnecessary because it is used more than it is needed. In design principles, redundancy is used for maintaining the performance of the system in the event of any failure. This concept is used significantly in digital message transmission in Electronics and Communication Engineering. The same messages are sent repeatedly and redundantly so as to ensure that the message is received from the transmitter to the receiver end. There are four types of redundancy approaches stated as follows:

(I) Diverse redundancy – In this type of redundancy, multiple elements of different types are used so as to maintain the system performance in event of any failure. For example, some innovative products and decorative items have two different projections or focus when viewed from different sides.

This is also done so that, in case the product gets damaged from one end, the other end of the product still fulfils the function.

(II) Homogeneous redundancy – In this case, a single element is used multiple times so as to maintain redundancy. This is applicable in the case of digital message transmission systems as discussed above, where the same message is transmitted multiple times so as to ensure that the receiver receives the message.

(III) Active redundancy – It is the application of redundant elements at all times. For example, multiple pillars for a roof. The main pillars in a small roof are the ones which are at the edges, but some extra pillars are added as redundancy for the support system to the roof in case of a collapse.

(IV) Passive redundancy – It is used whenever the active redundancy fails. For example, a spare tire present in a car is used in case of a flat tire or a puncture.

In succession, sometimes a concept of *Weakest Link* is provided deliberately in a design. The deliberate use of a weak element that will fail is provided in order to protect other elements in the system from damage. Interestingly while making prototypes for a product, the principle of *Scaling Fallacy* is extremely important to be observed. This principle is also known as the *Cube Law* or *Law of Sizes*. It states that, in general, it is a tendency to assume that a system that works on one scale will also work on a smaller or larger scale. However, this may or may not be true in every case. There are two main types of assumptions which hinder the scaling of any object or product. Load Assumptions occur when design and stresses on the base design are scaled equally, which may or may not be true. For example, while prototyping a missile, the scaled-down version might show success, but in actual practice, the missile launch may turn out to be a failure. Interaction Assumptions occur when interaction with design is considered invariant for any scaling, which may or may not be true. For example, while creating the models for a building and proposing a scaled-down version of the sizes and its structure, evacuation constraints in case of an emergency are sometimes overlooked. This is subjected to Interaction assumption of the prototype created.

1.6   WHY DESIGN AND TECHNOLOGY NEED TO GO HAND IN HAND

As is well introduced, technology is generated or created in order to ease the functioning of our day-to-day life. The evolution of technology has not only eased the lives of people but has also fostered creativity in the aesthetical launching of products.

However, it should be noted that design works in the front end of the system whereas technology works in the backend. Design refers to the interaction of the products by its users or customers who will then rely on the technology. If a design fails to convince the users, then even the best technology can be of no help. Thus, through this book, the detailed design process of evolution or re-design issues of products is discussed and then technology is introduced to make the product more sustainable in nature.

EXERCISES

1. Make a small presentation on the most important technology where design has brought a change in day-to-day life. Identify the different features of the technology which has left a huge impact on you and how the design intervention in the technology is addressed.

2. While working in the banking sector, privacy is extremely important to be honoured. Pick up at least ten examples where by using universal principles of design, the day-to-day load of banking personnel can be reduced. Compare your design ideations generated with both Direct Observation & Analysis and Indirect Observation & Analysis methods.

3. Reviews of a movie are always critical. Design two different questionnaires where the user can be interviewed taking the Hawthorne effect and Halo effect into consideration. Plot the responses on the empathy map and deduce your conclusions based on both the expectation effects taken separately.

4. Mobile Phones and Pagers were launched almost at the same time frame by different competitor companies. However, pager as a product though aesthetic and attractive lost its market share and was subsequently eliminated from the market. Identify the key design principles which were overlooked that led to the collapse of pager as a product.

5. Identify a customer who wants to buy a new Laptop. Using the Journey map method, provide the customer with the best choices available in terms of technology and other design principles associated with it.

6. Pick up the icon of any one of your favourite brands of accessories. The accessory can be watches, make-up kits, trending shoes, or any object close to your heart. Try finding out the reasons why the particular colour scheme is adopted while designing the icon. What happens when the icon is redrawn with cool or warm colours? Would you suggest any other colour combination for the icon? Explain using design principles.

7. Toys always draw major attention among users. It is extremely important that while the toys are affordable and cost-effective, they should also be attractive enough. Pick up five popular toys which infants use in the age group of 0–2 years. Identify what is the different mimicry techniques included in them. Also, find out what other design principles can be used to improve their outreach to infants.

8. Context is a very crucial concept in design. The design solutions for the same product can be different for different contexts of users considered. Take the case of a mobile phone. Identify the user needs of a mobile phone in three brackets of users: 18–45 years age bracket, 45–60 years age bracket and 60–75 years age bracket. Using the Direct Observation and Analysis method and using questionnaires and interview techniques, identify the detailed user needs for a mobile phone. How will you chunk the data received as a part of the interview process? Further, to map the qualitative details on a quantitative scale, you may use a Likert scale. Also justify, why the range of the Likert scale is so chosen.

9. Visit any organization, company or industry around you. Identify the happiness index of the workers by conducting an online survey. You may choose qualitative or quantitative questions to identify the happiness index. Now, by using a user-centric design approach, what design solution can be proposed to improve the happiness index of the organization? Repeat the same objective with a competent company or industry as observed before, using the design thinking approach. Compare the results of the two approaches in terms of qualitative and quantitative analysis.
10. Design a simple Mobile Stand for purpose of rooms and watching films at a convenience, including travel. What design principles have you taken into consideration? The user needs may be collected through the indirect observation and analysis approach.

REFERENCES

1. W. Lidwell, K. Holden, J. Butler, *Universal Principles of Design*, Rockport Publications, 2003.
2. S. A. Kadir, M. Jamaludin, "Universal Design as a Significant Component for Sustainable Life and Social Development", *Procedia – Social and Behavioral Sciences*, Vol. 85, pp. 179–190, 2013.
3. R. C. Atkinson, R. M. Shiffrin, "Human Memory: A Proposed System and Its Control Processes", In *Spence, K.W. and Spence, J.T. The Psychology of Learning and Motivation*, New York: Academic Press, Vol. 2, pp. 89–195, 1968.
4. G. M. Fix, B. Kim, M. A. Ruben, M. B. McCullough, "Direct Observation Methods: A Practical Guide for Health Researchers", *PEC Innovation*, Vol. 1, pp. 1–7, 2022.
5. M. T. Anguera, M. Portell, S. Chacón-Moscoso, S. Sanduvete-Chaves, "Indirect Observation in Everyday Contexts: Concepts and Methodological Guidelines within a Mixed Methods Framework", *Frontiers in Psychology*, Vol. 9, pp. 1–20, 2018.
6. R. M. Dios, M. A. Jimenez, "Polar Coordinate Analysis of Relationships with Teammates, Areas of Pitch, and Dynamic Play in Soccer: A Study of Xabi Alonso", *Frontiers in Psychology*, Vol. 9, pp. 1–12, 2018.
7. Robert S. Feldman, *Understanding Psychology*, McGraw Hill Education, 2015.
8. https://www.ideou.com/pages/design-thinking
9. Tim Brown, "Design Thinking", *Harvard Business Review*, June 2008.
10. Gerd Waloszek, "Introduction to Design Thinking", 2012.
11. https://www.interaction-design.org/literature/article/5-stages-in-the-design-thinking-process
12. S. Pande, A. Kenjale, A. Mathur, P. D. A. Kumar, B. Mukherjee, "Re-design of the Walking Stock for the Elderly using Design Thinking in Indian Context", *Innovative Product Design and Intelligent Manufacturing Systems*, Springer, Singapore, pp. 29–39, 2020.
13. https://careerfoundry.com/en/blog/ux-design/what-is-design-thinking-everything-you-need-to-know-to-get-started/
14. Nigel Cross, *Design Thinking: Understanding How Designers Think and Work*, Berg Publishers, 2011.
15. Tim Brown, Clayton M. Christensen, Indra Nooyi, Vijay Govindarajan, "On Design Thinking", *Harvard Business Review*, 2020.
16. Bill Burnett, Dave Evans, *Designing Your Life: How to Build a Well-lived, Joyful Life*, Knopf, 2016.

17. Michael Lewrick, Patrick Link, Larry Leifer, *The Design Thinking Playbook: Mindful Digital Transformation of Teams, Products, Services Businesses and Ecosystems*, Wiley, 2018.
18. J. Robert Rossman, Mathew D. Duerden, B. Joseph Pine II, *Design Experiences*, Columbia Business School Publishing, 2019.
19. P. Mukhopadhyay, *Ergonomics for the Layman*, CRC Press, 2019.
20. D. Chakrabarti, *Indian Anthropometric Dimensions for Ergonomic Design Practice*, National Institute of Design, 1997.
21. https://biologydictionary.net/anthropometry/

In a recent study in [4], it is claimed that about 80% of the elderly shall be present in low- and middle-income countries. This lays an emphasis on the needs that should be met to support such a huge chunk of geriatric or the elderly population. Especially keeping in mind that the major chunk of the elderly comes from low- and middle-income groups of countries pushes the need for products pertaining to the elderly as more economically affordable as possible. The various key point indicators for introspection of any design element for the elderly include Financial Security, Personal Security, Mental health, Health care and Self-Actualization. This is supported by Maslow's law of basic needs. Maslow's hierarchy of needs is shown in Figure 2.1.

The first stage of the law states that physiological needs include the needs of physical existence like food, water, breathing and excretion. All the needs which allow the physical self of a human body to exist fall under this category. The safety needs include safety for employment, financial safety, safety of resources, etc. A person always plans out certain defensible space around him or her, which is governed by the safe zone. These safety measures can be physiological safety needs or even mental safety needs, which is why safety for monetary needs in case of some bad times is also accounted for in this category. Many times, people plan out purchasing cars or houses under the Easy Monthly Instalment scheme (EMI) through bank loans. The loans are also planned out so as to develop the capability to bear the monthly EMI instalment to be paid for around 20 or 30 years. This falls under the Safety head of the law.

Love and Belonging connect one to family values and society through friendship, peer groups, etc. The near and dear ones, the emotional connection with peer group and friends and the social circle that one person maintains all fall under this category.

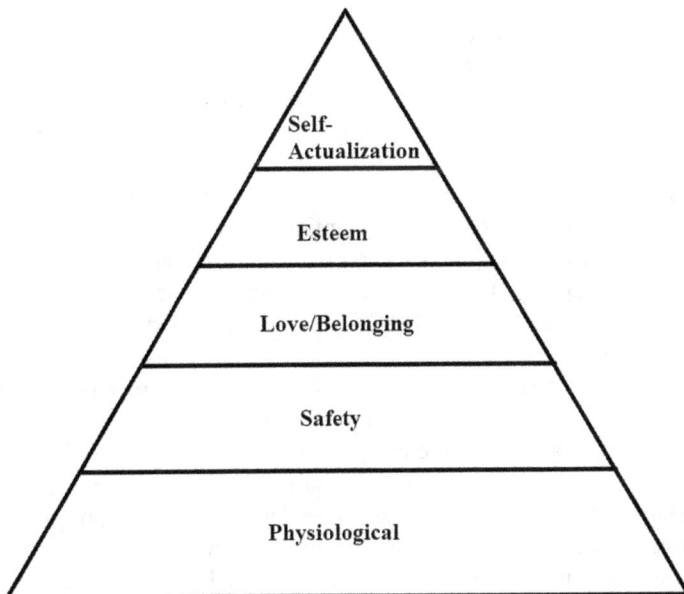

FIGURE 2.1 Maslow's law of hierarchy of needs.

The attributes of a human as a social animal refer to this stage of Maslow's law. The various ethics, manners, grooming, value education and teachings of an individual are glorified and utilised under the Love and belonging section. Esteem refers to the capability to behold your self-respect, confidence and your sense of achievement. It also refers to earningrespect from others and extending respect to others at this stage. It is one's deeds and work ethic that earn the respect of others. Self-esteem or self-respect of an individual also comes under the category of Esteem according to Maslow's law. The final stage of the law is Self-Actualization, which means one's morality, creativity, accepting facts and then rendering decisions based on the self-consciousness achieved. This stage of Self-Actualization is generally achieved after an individual has spent a long tenure of life and interacted with different kinds of people and faced different circumstances and situations in life. The situation of life can be both harmonious and discomfort. It is only after passing through the extremities of life that a person achieves the stage of Self-Actualization, which only comes through experience handling.

Thus, based on Maslow's law of hierarchy, it is essential to observe that most elderly users attain the stage of self-actualization and, thus, they do not like to remain dependent on anyone for their needs. To augment this, a major conclusion of the research findings in [4] emphasizes the need for an improved healthcare system, which needs to be extended to the elderly. The healthcare system needs to include innovation in healthcare products at affordable and economic prices to improve the self-confidence level among elderly citizens.

2.1 DAILY NEED PRODUCTS FOR DESIGN AND TECHNOLOGY INTERVENTION

In conjunction to the needs of the elderly, various healthcare products have been researched and innovated. The gamut of such products includes shower chairs with backrests, portable conversion commodes, wheelchairs with writing boards, air pump mattresses for bed sores, improved walking sticks, etc. The products can be broadly classified into two parts: one based on direct use and, second, indirect use. Indirect use covers such equipment which is used to monitor the daily routine habits of the elderly through a secondary or tertiary user. For example, medical equipment like portable ECG machines for geriatrics, improved blood pressure monitoring machines, etc. Though most of these equipment are improved for providing comfort to the user, they are used by Doctors (Secondary or tertiary users in the scenario with respect to the elderly) or medical practitioners. Direct-use products are the ones which are used by the elderly themselves. This includes improved commodes, mattresses, walking sticks, umbrellas, walkers, spectacles, etc.

The washroom is a place where every individual like to have their own personal space, peace of mind and independence. Seniors or the elderly, physically disabled people and individuals with mobility issues need a bit of extra support when it comes to answering nature's call or taking a nice, warm shower. It could just be a case of needing something to hold onto while getting up or being uncomfortable using a traditional Indian toilet because of knee problems. In such cases, there is a wide range of bathroom and toilet support products that serve the purpose and allow seniors to continue doing their daily routines in a hassle-free manner. For example, shower chair with a basket, toilet squat tool and wall-mounted India Commode conversion, etc.

Similarly, people with mobility issues (due to old age, injury or disability) find it extremely difficult to get up and sit upright without support. Some might need protection to avoid rolling off the bed in their sleep. There is a variety of bedroom accessories and support products that can help keep them stay comfortable, cosy and safe. Some of the innovative products include bedside rails, bed comfort wedges, etc. Similarly, a lot of other general well-being products for the elderly are also available in the market. These products or a catalogue of some of the products can be referred from [6, 7]. The mentioned products are a few among the many products available in the market. In fact, with the advent of e-commerce and websites, most of these products can be purchased through online mode also.

Interestingly, the researchers in the research article [5] investigated a detailed study on elderly consumers' new roles and related implications for business strategies, from a consumer behaviour perspective. Results present a structured classification of the most prominent streams of research by highlighting five promising changes (5Cs): changes in elderly consumers' roles in markets and societies, changes in self-care resulting in fashion purchases and cosmetic surgery, changes in elderly consumers' expenditures on specifically designed products and services, changes in the perception of risks resulting in preferences for either extremely prudent or hazardous behaviours and changes in general elderly characteristics due to the so-called ageless society.

Various previous studies show that with old age, expenses for food, transportation, personal care and clothing decline, while medical expenses and, in a linear way, those for heating increase. Elderly consumers need different products and services from the rest of the population; thus, everything should be redesigned, in shape and size, to adapt to the main physiological changes that characterise them as a population (reduced visual or auditory capacity, decreased motor skills). Moreover, to allow the elderly to be self-sufficient in carrying out their main daily tasks, such as dressing and washing, it is necessary to create optimal ergonomic conditions in their environment.

The new 'ageless society' requires new services aimed to sustain active ageing, such as plastic surgeries and other services aimed at enhancing body image. Thus, business efforts should be directed to ethically sustain such practices when positively related to the achievement of a positive mental state dictated by the desire to thin the boundary between chronological and cognitive age, such as in the case of plastic surgery. Considering the current scenario, it is very important to recognise interventions in the active ageing of the population, to provide elderly people with means to be able to counteract the effects of ageing and promote the adoption of healthier behaviours. Thus, these needs of the elderly should be addressed by the corporate sector so as to facilitate the concept of the new 'ageless society'.

While addressing the basic needs of the elderly, two important products play a significant role in the day-to-day life of any elderly. One is the walking stick and the other is the umbrella. In India, using a walking stick or umbrella as a major support system is a very common practice among the elderly. While these two products not only rely on a major physical support system, they also play a major role to improve self-actualization based on Maslow's law. While the market offers a huge choice of various improved designs for such users, it is equally important to address innovative solutions to these products so that their usability is leveraged.

Through this book, a detailed journey of design to technology incubation of the product lifecycle is discussed. It is important to note that the various parts of the products are essential to understand before the implementation of the design principles to improve and integrate the technology factors. Also, to clearly identify the role of ergonomics and anthropometry details, various relevant parameters are assessed so as to improve the design concepts and make them more user-friendly along with aesthetics. As has been well discussed, the products, especially for the elderly, also have an emotional connection. Thus, the design should also take into consideration the emotional quotient of the elderly associated with the products used by them.

2.2 INTRODUCTION TO THE WALKING STICK FOR THE ELDERLY AND ITS PARTS

It is a product used to facilitate walking, and for clarity of definition, it is a product used for a very personal or non-medical usage; used by the elderly for support, relief and confidence while walking. Those with medical conditions like knee injuries or ligament injuries, etc. also need a walking stick to assist in walking with confidence.

It can either be made of wood or light metal such as aluminium. The wooden sticks usually have a handle which cannot easily have multipoint tips. Usually, the grips are cylindrical in shape meant for holding the walking stick erectly. Wooden walking sticks in the native Indian language are also known as 'Laathi'.

Aluminium walking sticks can either have a flat or 'swan-neck' top and have the advantages of being adjustable in length and are able to have multiple points, for example, a tripod. It has been observed that wooden walking sticks are very popular in the rural or countryside areas of the country, while aluminium walking sticks are prevalent in the urban sector. This is because the wooden sticks are simplistic in design and relatively economical than the aluminium body-based counterparts.

The basic parts of a walking stick are detailed below:

Hand grips: This part is needed for tactile feedback and a good grip by the user. This is the uppermost part of a walking stick. A variety of styles and sizes are available. The type of hand grip prescribed or used depends on two important factors: first, the comfort of the patient or the user and, second, the grip's ability to provide adequate surface area to allow effective transfer of weight from the upper extremity to the floor. The various handles used for design of the walking sticks are depicted in Figure 2.2.

From Figure 2.2, it can be observed that the bed-type handle is mostly applicable in conventional wooden type walking sticks. This kind of handle is also used in conventional umbrellas. It provides a larger surface area to improve the grip and hold by the user, especially the elderly. The standard type of handle is mostly used with aluminium or metal body-type walking sticks. This is mostly used to ensure that the force executed on the palm is minimal. This kind of handle also reduces the pain points on the palm or the hand. The circular type of handle is mostly used in aesthetic forms of

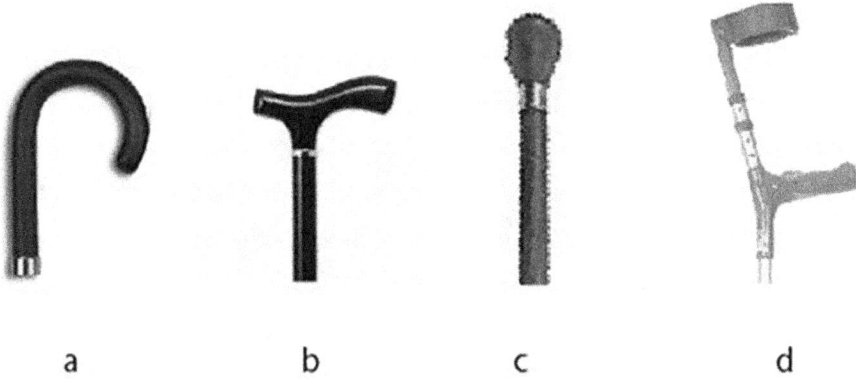

a   b   c   d

FIGURE 2.2 The various types of handles. (a) Bend (b) Standard (c) Circular (d) Handle for the handicapped.

walking sticks. Along with an appropriate colour scheme, the circular handle has been depicted as mostly a royal or chauvinistic personality-based person carrying a walking stick. These grips or handles are also popular while travelling from marsh landscapes or meadows. The handle for the handicapped is essentially extended up to the hand or the elbow so as to facilitate a better grip on the complete arm section for a physically handicapped person.

The various canes under consideration by usage in the market are discussed as follows:

Standard canes: Standard canes are lightweight and cost economical. They fit into the budget of the most socio-economic section of the society. The lengths of the wooden standard canes are mostly custom fitted to the specific patient, while the aluminium standard canes have pins for length adjustment so there is no need for custom fitting. Though so, the standard wooden walking sticks remain a popular choice for the elderly in general.

Offset canes: These canes are usually made from aluminium or some light metal and the lengths are also adjustable with no need for custom fittings. These canes allow the patient's weight to be displaced over the shaft of the cane. Through this way, the weight adjustment is achieved successfully. Further, the aim of metal- or aluminium-based body of the walking stick is to ensure that the stick remains light weighted.

Quadripod (Quad) cane: This is a four-legged cane usually made of aluminium. This cane permits more weight bearing, increases the base of support and provides more stability, in terms of walking, for the patient. It can also stand by itself freeing the patient to use his or her hands. This remains a popular choice among those users or elderly who have an acute knee problem or have undergone some surgery or medical treatment, generally in the knee section and now need a strong support system to walk independently without intervention of any other individual.

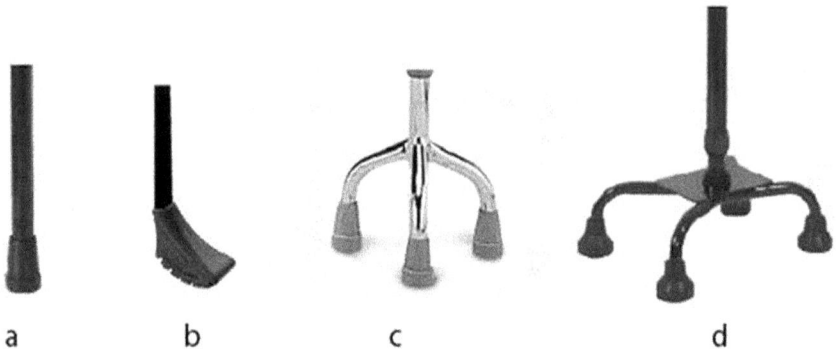

FIGURE 2.3 The various types of bases of the walking sticks. (a) Standard Single base (b) Bend base (c) Tri base junction (d) Quad base leg or junction.

> *Strong bases:* For providing a balance of weight and provide traction between the surface and device, a strong base is always recommended. There are various designs available in the market with different types of hand grips and bases. The most popular bases are shown in Figure 2.3.

The standard single base is part of the simplistic wooden stick, also called a standard cane or the 'Laathi'. The base may be added with an additional rubber grooving to provide extra stability or grip between the stick and the ground so that the user may feel more confident about using it. The bend base adds to a more contact area between the rest of the body of the walking stick and the ground, thus, displacing the weight more efficiently using the stick.

The tri-base junction and the quad-base junction both provide mechanical stability to the walking stick. By mechanical stability, it means that the sticks can be left to stand alone and would not roll or fall when left abandoned. The difference mostly lies in the surface area of the base in contact with the ground for the two bases. Users often find the quad base as a much stronger support than the tri-base junction. In fact for medical patients, a certain sense of insecurity always persists after some knee surgery or knee replacements. In such cases, the patients feel that the walking stick should be stable itself and standalone. Only when the walking stick is standalone can it actually bear the weight of the patient or the user. Thus, the three- and four-legged base becomes a popular choice in such a scenario. A gasket of rubber coating is further provided so as to increase the grip between the walking stick and the ground plane. In case, there is a tilt or bend observed while the user walks, the strong base provided a rigid support not allowing the patient or the user to fall.

2.3   PROBLEMS IN USING A WALKING STICK FOR THE ELDERLY

Finding out feasible and efficient solutions for elderly related products has always been a major subject of interest not only for designers but also for engineers. A major branch of design now emphasizes emotions connected to the products of interest to human beings. It is interesting to note that an umbrella and a stick are quite important and essential tools which also connect emotionally to the elderly in India. Whereas an umbrella is ought to provide shelter from rainfall and heat from sunrays, it is also a major tool of support and balance during a walk for an elderly.

A similar functionality of the stick for balancing oneself during a walk is well known. Further, analysing the varied nature of the Indian climate from scanty rainfall in deserts in Western India to heavy rainfall in eastern and southern parts of the country, the lacunas in the design issues of the conventional umbrellas and sticks specifically catering to the needs of elderly need to be assessed in detail. Apart from ergonomic issues and power outages at night during monsoon season in developing countries like India and Southeast Asia, the physiological parameters of an average elderly in India shall also be taken into account while proposing any re-design of the products.

Some statistics that direct us towards a general idea of the background to the re-design of the products selected for the scenario in the Indian context are as follows:

1. 65% of rural India has electricity as their primary source of energy [8]. Electricity generation comes from various sources like hydel power, thermal power and nuclear power plants in the country. However, providing consistent and uninterrupted power to such a huge landmass and demography remains a challenge.
2. Around 20% use a cane for walking [8]. By cane, the standard wooden walking stick is referred. Canes are relatively simple in design and easy in terms of availability. Further, canes are economical in terms of affordability.
3. Umbrellas are used even in summers and during days to protect oneself from the sun. Since the Indian subcontinent is surrounded by the sea and oceans, there is bright sunshine and also temperatures rise in the plains of the country. In such scenarios also, umbrellas are typically used for protection from sun rays. Umbrellas are not just restricted to use during rainy seasons or monsoons only. In fact, from the reference [9], the emotional attachment of the elderly in the Indian subcontinent is extensive, to their personal belongings and the products they use, especially the umbrella. The umbrellas are sometimes inverted and also used as a walking aid, providing balance during movements or a walk.

The problems faced by all the elderly are stated as follows:

1. There is a loss of efficiency in performing any action [10–12]. Some actions and their losses, in percentages, are as mentioned:
    i. locomotion 35%
    ii. bending 28%
    iii. remembering 22%
    iv. twisting 20%
    v. reaching 18%
    vi. hearing 13%
    vii. grasping 13%

viii. visibility 11%
   ix. eating 5%
2. Sensation and perception diminish with age.
3. Ocular motility is impaired with age.
4. Field of vision becomes smaller.
5. Presbyopia is common resulting in reduced accommodation.
6. Less light passes through lenses.
7. Cataracts, floaters and glaring are common.
8. Yellowing of the lenses occurs.

With the above-stated limitations of the elderly, products like umbrella and walking stick re-design need to be addressed with a better emotional connect. This can be done using both user-centric design and design thinking approach. However, since providing technology-based solutions to products to leverage the overall quality of life of the elderly remains prima facie, the design thinking approach is better suited to provide rational solutions.

2.4   INTRODUCTION TO UMBRELLA FOR ELDERLY AND ITS PARTS

Umbrella is a product designed to protect against the sun or rain. An umbrella is a folding canopy supported by wooden or metal ribs, which is usually mounted on a metal, wooden or plastic pole. An umbrella may also be called a brolly (UK slang), parapluie (nineteenth century, French origin), rain shade, gamp (British, informal, dated), or bumbershoot (rare, facetious American slang). When used for snow, it is called a paraneige. The term 'umbrella' is traditionally used when protecting oneself from rain, with parasol used when protecting oneself from sunlight, though the terms continue to be used interchangeably.

Umbrellas can be divided into two categories, namely fully collapsible umbrellas, in which the metal pole supporting the canopy retracts, making the umbrella small enough to fit in a handbag and non-collapsible umbrellas, in which the support pole cannot retract and only the canopy can be collapsed. Another distinction can be made between manually operated umbrellas and spring-loaded automatic umbrellas, which spring open at the press of a button. Usually, a button is provided next to the handle of the umbrella.

The basic parts of an Umbrella are:

1. *Canopy:* The umbrella canopy is the fabric used to cover the umbrella ribs. Fabrics used include polyester, nylon, cotton or plastic, with the most common fabric being polyester as it is waterproof, durable and retain its shape whether wet or dry. A teflon coating can also be applied to the canopy to further improve the waterproof properties and to aid the removal of stains from the surface and the metallic body frame.
2. *Cap:* The umbrella cap is found at the top of the umbrella just above the canopy. The function of the cap is to ensure that the water flows onto the canopy and not down the umbrella shaft.

3. *Ferrule:* Usually made of metal, ferrule adds strength to the end of the umbrella that extends beyond the canopy. It prevents the end of the umbrella from splitting or wearing, and is an important aspect that prolongs the life of the umbrella.

4. *Frame:* The umbrella frame encompasses ribs and stretchers that form a structure upon which the canopy is sewn. Umbrella frames are usually made of metal, fibreglass or plastic and can have between 6 and 24 ribs, with 8 ribs being the most common. The frame has fulcrum points and joints that allow the frame to collapse into the closed position to allow the umbrella to be carried more easily. Further, the metal is so chosen which reduces the weight of the umbrella and does not add to any pain points on the shoulder, wrist, or hand.

5. *Handle:* The handle of an umbrella, used to hold the umbrella, is found at the base of the shaft and made of wood or plastic. Handles are usually straight or curved and vary in size greatly. Various types of handle geometries and shapes are available in the market. Some of the popular shapes of the umbrella handle are similar to the handle of the walking stick as depicted in Figure 2.2, except for the handle for the handicapped.

6. *Notch:* The notch of the umbrella is the ring towards the top of the umbrella shaft where all the ribs come together and are held in place. It can be seen just below the canopy when looking from the underside of the umbrella and is usually made of plastic or metal.

7. *Runner:* The runner of an umbrella is the ring that moves up and down the shaft that allows the ribs to fully extend as the umbrella is opened and contract when the umbrella is closed.

8. *Shaft:* The umbrella shaft connects the frame to the handle. It is usually made out of wood, metal, fibreglass or plastic. Stick umbrellas have a continuous solid shaft thus offering a strong core to the umbrella. These are usually cylindrical in shape.

9. *Stretcher:* The stretcher is made of metal, fibreglass or plastic. They connect the ribs to the runner of the umbrella. As the runner moves up the shaft, it causes the stretchers to extend outwards. They then place pressure on (or stretch) the umbrella ribs so that they stand firm and form the arc of the open umbrella.

10. *Top spring:* The top spring of the umbrella is located towards the top of the shaft. It is a feature of manual-open umbrellas and is the triangular-shaped piece of metal that protrudes from the shaft to hold the runner in place when the umbrella is open. It is flat on the top edge to allow the runner to rest on top easily and is connected to a spring to allow it to hold firm. When the umbrella is to be closed, the top spring is pressed towards the shaft to allow the runner to pass by so the umbrella can be closed.

The various parts of the umbrella are clearly depicted in Figure 2.4. Figure 2.4a shows all the interior parts of the umbrella, whereas Figure 2.4b shows the outer periphery of the umbrella. The arc of the umbrella can be clearly observed in the figures. The basic parts remain invariant in all types of umbrellas. What adds to the

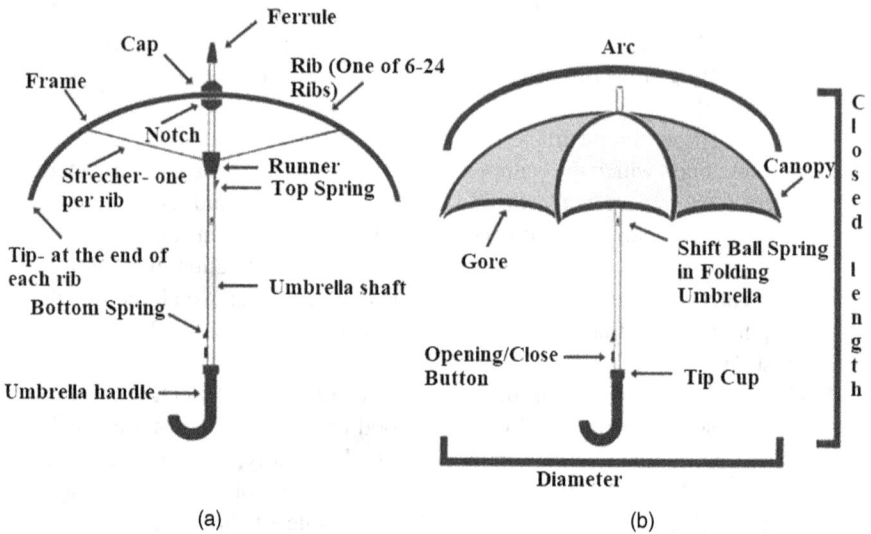

FIGURE 2.4 The various parts of the umbrella. (a) Inner parts (b) Outer periphery.

innovation in the design is a factor of portability. The umbrella should be portable enough so as to be carried conveniently. The trade-off in the design is that while it is expected that the umbrella should be folded to the minimum size so that it becomes easily portable, the canopy should be large enough so as to provide an adequate coverage during rains or protection from sun rays.

This also lays equal emphasis on the choice of the cloth material for the umbrella so that it is not tampered or torn while folding, and it becomes small enough to facilitate portability. To augment, the contacts should not stain or rust over the course of time, keeping the weight of the umbrella low enough so that it does not become a burden for the user or any elderly to carry.

The different types of umbrellas used by the elderly based on the opening of umbrellas are listed as below:

(A) *Manual umbrella:* A manual umbrella is one that is opened and closed by hand without any automatic mechanism. To open the umbrella, the runner is manually lifted along the shaft and is held in place towards the top of the shaft by the top spring. The umbrella can then be manually lowered by pressing the top spring and sliding the runner down the shaft until the umbrella is fully closed. Manual umbrellas are sometimes also called conventional umbrellas.

(B) *Automatic umbrella:* An automatic umbrella is one that opens automatically or opens and closes automatically, activated by a button. When the button is pressed, the umbrella will open, which is caused by springs or pressure from the ribs and stretchers.

2.5 PROBLEMS IN USING AN UMBRELLA FOR ELDERLY AND ITS PARTS

Keeping the points discussed in Section 2.3, the issues of the usage of umbrellas by the elderly are also attributed to body discomfort mapping. Similar cases as applicable to using a walking stick limit the lifting of the umbrella and using them for longer duration either erect or vertically or by using a support on the shoulder. During monsoons or rains, the cloth of the umbrella on account of becoming wet also makes the umbrella heavy, which may further add to the body discomfort or pain points. These all need to be taken into account while re-design issues are addressed.

Umbrella is used by the elderly not just for protection from rain but also for protection from sunrays. As was already mentioned before, conventional umbrellas are also used sometimes as walking sticks to aid during a walk. This also throws light on the fact, based on the emotional connect of the elderly or users, umbrellas become a part of some necessary products as a personal belonging. Thus, design inputs that empathise with the elderly or the user become a prima facie.

In case the umbrellas are big and bulky to cover the maximum area for protection, the weight becomes a major issue. Further, use of umbrellas in rainy season during the dark purports a challenge of using two products, namely an umbrella and a torch, which adds to the cognitive load of the elderly. Thus, design solutions to both walking stick and umbrellas become a necessity specifically catering to the elderly requirements.

2.6 SCOPE OF THE WORK

While it is evident that the products, namely the umbrella and walking stick, need intervention so as to make them user-friendly from the elderly perspective, some other areas of technology intervention are also needed to enhance the overall product life and usability. For this purpose, certain challenges and problems have also been identified.

There exist problems related to the environment when the user is present in a scenario as referred from [13]. When the elderly use a walking stick or umbrella in the dark, which is common due to the power cuts that take place on rainy days or monsoon season, a lot of different issues are reported. Some of them are listed as follows:

1. Location of the lights in the products of the elderly at intervals causes alternate bright and dark spots. This is because the lights on such products may be present at some random locations or the series connections between the lights on the products may have some gap between two bulbs or any two light sources, thus leading to alternate bright and dark spots. This may cause a hazy or unclear vision for the elderly while using the products and may lead to a possible accident.

2. Absence of ambient light during night-time in rural areas is a serious problem. It is quite possible that there might be no proper street lighting in such remote areas. Further, looking at the remoteness of such areas, lack of lighting in products may affect visibility, which may cause overlooking of some insects or animals, leading to possible accidents.

3. Dimmer lights due to improper functioning of equipment. This may also be attributed to discharged batteries. Battery life and the type of light source used affect the mobility of such equipment significantly. Even though the battery life may not be extendable based on the commercially available solutions, the product life can be improved by looking for lighting solutions like Light Emitting Diodes (LEDs), which consume comparatively less power but offer a high luminance radiating out from them.

The genesis of this case study is attributed to understanding of a course on Product Design in Electronics for design students emphasizing how designers and Electronics/ Electrical Engineers can join hands in improving the quality of life for the present socio-economic structure of a middle-class citizen. The most of the research findings are delved within the Indian Context.

This book describes the walk-through of how electronics, when integrated into simple products like umbrellas and sticks, can ease the life of an elderly, keeping in mind the various physiological and ergonomic parameters in the Indian context [14]. Technical solutions to provide lightweight and electronics-equipped umbrellas and walking sticks shall be made with better ergonomic features like grip analysis, material for fabrication, etc.

Once the design and fabrication of the prototype are done, a user study to measure the user response of the prototype is done. This user study is done to judge the products so developed both qualitatively and quantitatively.

Hence, the identified user needs are as follows:

1. Higher intensity of light is needed to aid low vision, without glare and must be uniform
2. Easy to use
3. Give cane-like usage facility
4. Design for emergencies

This book, thus, presents a perfect amalgamation of the Design and Technology for Product or User Centric Design Principles or Design Thinking approach that can be adopted.

Among some of the principles, *Path of Least Resistance* is of vital importance for such products. This principle is also known as *Performance Load* or *Principle of Least effort*. According to this principle, the greater the effort it takes to accomplish a task, the less likely it is that the task will be accomplished successfully. This means if it takes extra effort for completion or accomplishing a particular task, the likelihood that the task shall be successfully accomplished has very low chances. One would always like to accomplish a task with less effort. This can be understood by a simple mathematical example of algebra. One can simply write $2 + 2 = 4$, while teaching algebra to kindergarten students. However, if someone now writes variables as $x = 2$, $y = 2$ and $z = x + y = 4$, then this unnecessarily complicates a simple learning process. This should be avoided in general.

There are two types of load that a user undergoes in the path of least resistance. This is as mentioned below:

(A) First is the cognitive load, which states that it is the amount of mental activity in terms of perception, memory and problem solving that is required for accomplishing a task or a goal. For example, the number of instructions taken and executed at a time by any person depends on the capability of the individual to handle the cognitive load.

(B) The second is the kinematic load. It is the degree of physical activity required to accomplish a task or a goal. The effect of Kinematic load can be understood by the difference between the presence of staircases and an elevator or an escalator. Given the option to move from one floor in a building to the other floor or while travelling in a shopping mall, most people prefer to use an elevator or an escalator rather than climbing the staircases in order to reduce the kinematic load.

Thus, while designing products for an elderly, especially walking sticks or umbrellas, it should be kept in mind that neither the cognitive load nor the kinematic load should increase for the user.

It should also be understood that designs that help perform optimally are often not the same as the designs that people find most desirable. This is called the choice between *Performance versus Preference*. Further, this is also in tandem with the Satisficing Principle.

In the previous chapter, a detailed discussion of Hick's Law has been done. This is applicable in the decide option while performing any task. This law states that the time it takes to make a decision increases as the number of alternatives increases. While empowering the products for the elderly with technology-based solutions, it should always be kept in mind that with age, as pointed out before, the elderly cannot handle extensive cognitive load. Thus, technology-based solutions should not add to cognitive load on the users such that it leads to collapse in the use of the product based on Hick's Law.

Since the case studies under discussion include walking stick and umbrella for the elderly, a discussion on the Structural forms becomes important. Structural forms refer to various forms as applicable to support any load. This mainly has three forms, as stated below:

(i) *Mass structures* – This refers to the form solid structure as a function of weight, hardness, etc. These structures are not hollow; instead, they are made up of a complete material or of a combination of different materials called composite materials. Examples include bricks and stones.

(ii) *Frame structures* – These are elements which are joined to form a framework, for example, trusses. By definition, a truss is a structure that consists of two-force members only, where the members are organized so that the assemblage as a whole behaves as a single object. A planar truss is one where all members and nodes lie within a two-dimensional plane, while a space truss has members and nodes that extend into three dimensions. An example of a truss-based structure is the Howrah Bridge in Kolkata, India, or the Eiffel Tower in Paris, France.

(iii) *Shell structures* – These are the structures which are used to wrap around and contain certain volume inside it. For example, aeroplanes. Aeroplanes are sheet metal framework-based structures, which are kept hollow to incorporate the passengers inside it. Once the aircraft is sealed and locked from inside, it is ready for flying in the sky.

While ideating any object or for creation of visual perception of an object, three-dimensional projections are extremely important. There are many ways in design where the three-dimensional projections can be rendered on a paper which is two-dimensional in nature. Some of the techniques are as follows:

(a) Visual cues used are interpreted as Interposition. The overlapped object is perceived as farther away from overlapping object.

(b) Size can also be rendered on two-dimensional pages. Smaller Objects are perceived as farther away than larger object

(c) Similarly, elevation can be shown as a three-dimensional projection. Objects at higher elevation are perceived to be farther away than the nearer ones.

(d) To scale the linear perspective, converging ends of vertical lines are perceived to be farther away from diverging ends.

(e) Texture gradient can be shown in terms of density in any area under consideration. Areas of greater density are perceived to be farther away from areas of lesser density.

(f) Shading plays an important role in perception and to create three-dimensional projections. In shaded regions, light-shaded areas are perceived to be closest to light sources, and the shaded areas are considered farther away

(g) To create atmospheric or ambient perspective, blurs may be deployed. In multiple objects placed in a scenario, objects that are blue and blurry are perceived farther.

EXERCISES

1. In order to understand the design thinking process or user-centric design, visit a nearby old age home or elderly care centre. Collect the different walking sticks or walker from the elderly and put them in one place in a scattered manner. Now create a three-dimensional rendering of the walking sticks by sketching on a paper. You may change the scenario by interchanging the different sticks o swapping one from other and rendering the situation again on the paper.

2. Prepare a questionnaire to plot the happiness quotient of the elderly in your surroundings. Using Direct Observation and analysis through interviews and questionnaires, find out whether the elderly in your surroundings support

disengagement theory of ageing or activity theory of ageing. You may also use empathy maps to verify your findings.

3. Mobile Phones are extremely important for communication, especially for breaking the distance barrier. The elderly too use mobile phones; however, the way they look at mobile phones has a different perspective than teenagers or children in their adolescence age. Using Direct Observation and Analysis method, conduct a survey using Questionnaire and interview to find out the importance of mobile phones for different age groups. Plot the findings into empathy maps. What do you deduce from the empathy maps as far as the emotional connect of mobile phones with their respective users are concerned? What do you find out about the correlation of the same spanned over the various age groups?

4. Repeat the experiment as stated above again. However, this time, using Direct Observation and Analysis, find out what are the true needs of a mobile phone for the elderly. Propose your design solution of the interface of a simplistic mobile phone that suffices the purpose of the recordings collected by you in the survey. Also, repeat this experiment based on data collected from different sources using the methodology of Indirect Observation and Analysis. Do you find similarities in your recordings of the data collected? Would it also lead to the same ideation of concepts that you have rendered as a finding of the Direct Observation method?

5. Medical equipment have undergone extensive design iterations. One of the key equipment is weight measurement. In almost all places, a standard weighing machine is used which can be mechanical or digitized in nature showing accuracy up to the second or third place of decimal. Identify the different mechanisms of weight measurements that are used in hospitals or nursing homes for patients. Further, find out the reasons why the particular design solution for the weight measurement was proposed. Through this exercise, you will be able to backtrack the design process that the designers might have undergone to propose the modified product.

6. Identify the materials of cloth which are available in the market for commercial use on umbrellas. Consider the case of (A) bright sunshine, (B) extreme high rains in monsoons, (C) heavy downpour rains, (D) use by elderly and (E) moderate weather conditions. Which material do you suggest for the cloth under all the different conditions? If you need to choose a universal material catering to all the cases as stated above, which one would you prefer as a designer and why? Justify your reasons based on the various Universal Principles of Design.

7. Consider the different types of umbrellas that are available in the market. You need to choose from all manual and automatic. Identify, how the canopy design, joints and structure are modified from technology perspective for each of the products. Prepare a comparison table of each of the design and identify the strengths and weaknesses of each design.

8. Take the case of the base of a walking stick. It is interesting to note that the base of a walking stick either is one-legged or three-legged or four-legged. Why is two-legged walking stick not preferable? What are the possible

design bottlenecks which might hinder its commercialization? Identify the constraints with reference to the mechanical stability of the walking stick as well. Would you propose or ideate a two-legged walking stick for medical patients who have undergone knee replacement or knee-related surgeries? Why or why not?

9. In a recent case as flashed in the news, an undergraduate student of a prestigious university in India was flying to New Delhi from Geneva when a co-passenger behind him needed medical assistance. The passenger was a Type 1 diabetes patient and had forgotten his insulin pump at Moscow's Sheremetyevo International Airport during the security check. It had been five hours since his last insulin dose and his blood sugar had reached dangerous levels. Though the passenger was carrying insulin cartridges, he did not have the device needed for injecting it into his body. A doctor on the same flight, also a diabetes patient, had multiple injecting equipment at hand but none could match the passenger's insulin pen's diameter. What are the possible ways or ideations that you can generate of proposing a universal needle, which can fit into the insulin injections manufactured by any company?

10. Identify the different contexts where you can propose any user a solution of standard walking cane or offset canes. Which would you suggest from (A) an elderly who is 90 years old and needs support to walk and (B) a 40-year-old patient who has undergone a knee surgery?

REFERENCES

1. M. A. Makary, D. L. Segev, P. J. Pronovost, et al., "Frailty as a Predictor of Surgical Outcomes in Older Patients", *Journal of the American College of Surgeons*, Vol. 210, No. 6, pp. 901–908, 2010.
2. https://www.msdmanuals.com/en-in
3. Robert S. Feldman, *Understanding Psychology*, Mc Graw Hill Education, 2015.
4. Amelia Anggarawati Putri, Chairunnisa Niken Lestari, "The Ability to Meet the Elderly's Basic Needs for Healthy Ageing in Low- and Middle-Income Countries", *ICGH Conference Proceedings*, 2017.
5. G. Guido, M. M. Ugolini, A. Sestino, "Active Ageing of Elderly Consumers: Insights and Opportunities for Future Business Strategies", *SN Business and Economics, Springer*, Vol. 2, No. 8, pp. 1–24, 2022.
6. https://www.eldereaseindia.com/
7. https://www.seniority.in/
8. Aditya Ramji, Anmol Soni, Ritika Sehjpal, Saptarshi Das, Ritu Singh, "Rural energy access and inequalities: An analysis of NSS data from 1999-00 to 2009-10", The Energy and Resources Institute TERI-NFA Working Paper No. 4, December 2012.
9. S. T. Charles, L. L. Carstensen, "Social and Emotional Aging", *Annual Reviews of Psychology*, Vol. 61, pp. 383–409, 2010. doi: 10.1146/annurev.psych.093008.100448. PMID: 19575618; PMCID: PMC3950961.
10. Karl Kroemer, "Extra-Ordinary Ergonomics", *Chapter: Design for Aging*, Taylor & Francis.
11. Robin M. Daly, "Independent and Combined Effects of Exercise and Vitamin D on Muscle Morphology, Function and Falls in the Elderly", *Nutrients*, Vol. 2, pp. 1005–1017, 2010.

12. Sushma Tiwari, A.K. Sinha, K. Patwardhan, Sangeeta Gehlot, I. S. Gambhir, S. C. Mohapatra, "Prevalence of Health Problems Among Elderly: A Study in a Rural Population of Varanasi", *Indian Journal of Preventive and Social Medicine*, Vol. 41, pp. 226–230, 2010.
13. "Lighting your way to better vision", *IESNA Lighting Handbook*, 9th edition. Illuminating Engineering Society of North America, 2009.
14. "Lighting Solutions for Elderly Healthcare", Brochure of Derungs Licht AG.

3 Re-design Issues of the Walking Stick for the Elderly

Finding out feasible and efficient solutions for the elderly has always been a major subject of interest not only for designers but also for engineers. A major branch of design now emphasises emotions connected to the products of interest to human beings. It is interesting to note that a walking stick is quite an important and essential tool, which also connects emotionally to an elderly in India. A walking stick for balancing oneself during a walk is well known. However, the emotional attachment and the qualitative/quantitative analysis of the walking stick uses by the elderly, forced the genesis of the problem statement on the re-design issues of the walking stick in the Indian context. The proposed solutions of the walking stick encapsulate the relevant Anthropometric data of the Indian elderly augmented with technological solutions like Lighting Solutions and SMS with GPS location in case of an emergency situation. The aesthetics (in terms of colour of the stick and ideated designs) and common issues like stronger grip, footrest and capabilities of alarm system have also been taken into design consideration. A detailed user study for proposing the walking stick design using design thinking and the prototype of designed stick with technological incubation has been carried out. The results of the user study reveal considerable acceptance by the users. The walk-through in the re-design of the products discussed as case studies in the book is an outcome of two courses, namely Science in design and Design and Technology. Thus, students from both design and electronics department can participate in experimental validations.

3.1 ANTHROPOMETRY OF THE INDIAN ELDERLY

The first step to the redesign of the walking stick deals with the collection of relevant anthropometric data. There is a vast literature on Indian anthropometry available in [1]. The data in [1] covers the age bracket of 18–80 years collected from about 23 cities in India. Since the walking stick product caters to the need of the Indian elderly, initial reference to map the relevant data of the average Indian elderly (both men and women) is collected.

All standard practices as mentioned in [1] are followed for collecting the data of relevant anthropometric parameters. The procedures for the collection of the relevant anthropometric data need to be followed. Generally, the subject whose anthropometry data is to be measured is stripped up to the waist and should preferably be asked to wear shorts or Bermuda. In case of female, the data collection should be done in a private place where a female attendant should only take the measurements after taking prior approval from the subject. The permission is also to be taken from males. This is a part of ethical clearance, where any data to be collected from humans should

DOI: 10.1201/9781003414957-3

be done with due permission. The subjects should always be made aware of what kind of measurements shall be done with them and for what reasons. This eases the subject to cooperate with the measurement procedures.

The data of four identified dimensional measurements, presented herein, are collected from males and females of 60 years and more (with ratio, 1.03:1) from mixed occupational groups, engaged in a variety of activities at home, offices, educational organizations, agriculture, business, industries etc. from different locations of India.

Some of the locations covered during the study for the collection of the data are as follows:

1. Jabalpur (Madhya Pradesh)
2. Kadapa (Andhra Pradesh)
3. Ongole (Andhra Pradesh)

All three places are in India. Apart from this, the data is also collected from some other cities in different states of the country as well. A total of 81 male senior citizens and 78 female senior citizens are covered in the measurement process.

The whole survey was organised by visiting respective locations of the people like homes and workplaces. Prior permission from the subjects was taken as a part of the ethical clearance needed for the measurements of human subjects. Various dimensions were measured directly using the measuring tape and the usual standard procedures as mentioned in [1], which were recorded in data sheets followed by relevant statistical compilations.

For some measurements, the subjects were asked to be in required posture and movement ranges. For some other measurements, the subjects were allowed to follow their own comfortable postures and movement ranges.

The various limitations in the measurements observed are as follows:

- All dimensions were measured through usual practice. There may be a possibility of error in some measurements. A strong possibility of error is the parallax error during measurements. The error/displacement is caused in the apparent position of the object due to the viewing angle that is other than the angle that is perpendicular to the object. A common parallax error occurs in the chemistry laboratory. It is failure to read the volume of a liquid properly in a graduated cylinder or burette. This is also very common while performing the titration experiments or reading the volume of liquid from a pipette.
- Further, the measurements are subjected to the least count of the measuring device or technique deployed. In the science of measurement, the least count of a measuring instrument is the smallest value in the measured quantity that can be resolved on the instrument's scale. The least count is related to the precision of an instrument; an instrument that can measure smaller changes in a value relative to another instrument. Least count uncertainty is one of the sources of experimental error in measurements. Least count of a Vernier calliper is 0.02 mm and least count of a micrometre is 0.01 mm in general. Similarly, every equipment or device used for measurements in Anthropometry also has a certain least count which results in uncertainty leading to error while measuring.

3.1.1 STATISTICAL TREATMENT

The collected data is statistically studied to know dimensional variations of 159 people (81 male senior citizens and 78 female citizens). Tools like error, percentiles, mean and standard deviation are used to interpret them statistically.

Mean: Mean is the best measure of the central tendency of the distribution [1–4]. It simply refers to average. However, there are three types of mean which are of interest to the researchers in statistical treatment.

The Arithmetic Mean (AM) or average value of a score is the sum of all of the numbers divided by the number of numbers. This is statistically shown as $AM = \dfrac{\Sigma X}{N}$, where X = individual score, ΣX = Sum of the individual scores, N = number of subjects. For example, for a collection of five numbers namely 3, 4, 6, 8, 9, the arithmetic mean is given by $AM = \dfrac{3+4+6+8+9}{5} = 6$.

The Geometric mean is the average in terms of the products of the numbers. This is statistically shown as $GM = \sqrt[n]{\prod_{j=1}^{j=n} X_j} = \left(X_1 \times X_2 \times X_3 \times \ldots \ldots \times X_n \right)^{\frac{1}{n}}$, where X = individual score, $\prod X$ = Multiplication of individual scores, n = number of subjects. For example, for a collection of five numbers, namely 3, 4, 6, 8, 9, the geometric mean is given by $GM = \sqrt[5]{3 \times 4 \times 6 \times 8 \times 9} = 5.53264$.

The Harmonic Mean is an average which is useful for sets of numbers which are defined in relation to some unit, as in the case of speed (i.e., distance per unit of time). This is statistically shown as $HM = n \left(\sum_{j=1}^{n} X_j \right)^{-1}$, where X = individual score or value, j = the number of such values taken into consideration. As an example, harmonic mean of five values namely 4, 36, 45, 50, 75 is given as $\dfrac{5}{\frac{1}{4}+\frac{1}{36}+\frac{1}{45}+\frac{1}{50}+\frac{1}{75}} = \dfrac{5}{\frac{1}{3}} = 15$.

The Quadratic Mean or the Root over the Mean Square (RMS) is the absolute magnitude of a set of numbers or values which are measured. This is statistically defined as

$$QM \text{ or } RMS = \sqrt{\dfrac{\sum_{j}^{n}\left(X_j\right)^2}{n}} = \sqrt{\dfrac{X_1^2 + X_2^2 + X_3^2 +\ldots\ldots\ldots+ X_n^2}{n}} . \text{ For example, the}$$

RMS value of 2, 3, 4, 5, 6 is $\sqrt{\dfrac{2^2+3^2+4^2+5^2+6^2}{5}} = \sqrt{\dfrac{4+9+16+25+36}{5}} = 4.2426$.

It is of interest to note that for the same data set values, RMS (QM) $\geq$ AM $\geq$ GM $\geq$ HM, where the equality holds valid if the data sets are all the same values.

Another important parameter of interest is the Median. The median is the value separating the higher half from the lower half of a data sample, a population, or a probability distribution. For a data set, it may be thought of as 'the middle' value. The basic feature of the median in describing data compared to the mean (often simply described as the 'average') is that it is not skewed by a small proportion of extremely large or small values, and therefore provides a better representation of a 'typical' value.

Example 3.1

A certain measurement of the length of a pen using Vernier calliper measures 10 cm, 10.1 cm, 10.3 cm, 9.8 cm and 9.9 cm in five attempts. What are the Quadratic Mean, Arithmetic Mean, Geometric Mean and Harmonic mean of the measurements?

SOLUTION

$$QM \, or \, RMS = \sqrt{\frac{10^2 + 10.1^2 + 10.3^2 + 9.8^2 + 9.9^2}{5}}$$

$$= \sqrt{\frac{100 + 102.01 + 106.09 + 96.04 + 98.01}{5}}$$

$$= \sqrt{100.43} = 10.0215$$

$$AM = \frac{10 + 10.1 + 10.3 + 9.8 + 9.9}{5} = \frac{50.1}{5} = 10.02$$

$$GM = \sqrt[5]{10 \times 10.1 \times 10.3 \times 9.8 \times 9.9} = \sqrt[5]{100929.906} = 10.01853$$

$$HM = \frac{5}{\frac{1}{10} + \frac{1}{10.1} + \frac{1}{10.3} + \frac{1}{9.8} + \frac{1}{9.9}} = 10.017$$

Errors: The uncertainty in a measurement is called an *error*. It is the difference between the measured and the true values of a physical quantity. In contrast, *discrepancy* is merely the difference between the two measured values of a physical quantity. The different categories of errors are as stated below:

(A) *Constant errors* – If the same error is repeated every time in a series of observation, the error is said to be constant. This may occur due to faulty calibration of the scale of a measuring instrument. In order to minimise constant error, measurements are made with all possible different methods. The mean value so obtained is regarded as the true value.

(B) *Systematic errors* – These are those errors which occur according to a certain pattern or system. These errors occur due to known reasons and can be minimised by locating the source of the error. This is subcategorised into the following types:

(i) *Instrument errors* – This error occurs if an instrument is faulty or inaccurate. Either the interchange of two similar instruments or the use of different methods to measure the same quantity can be of some help in minimizing the instrumental errors.

(ii) *Personal errors* – Sometimes, errors in the measurements are due to the individual qualities of the experimenter itself. These errors arise due to lack of attentiveness, bad sight, habits or peculiarities of the observer. The parallax error comes under the personal error category. In order to eliminate personal error, the measurements are repeated by different observers.

(iii) *Errors due to external sources* – These errors are caused due to change in external conditions like pressure, temperature and wind. These errors can be minimised by keeping control over external conditions in which the experiment is performed.

(iv) *Errors due to internal sources or errors due to imperfection* – These errors are due to limitations of experimental arrangement. For example, the loss of energy due to radiation causes errors in calorimetric readings.

(C) *Gross errors* – These can be attributed to many reasons like improper setting of the measuring instrument, recording the observations incorrectly, not taken into account the sources of errors and their precautions and using some wrong value in calculations. These can be minimised only if the observer is very careful in his or her approach.

(D) *Random errors* – It is observed as a common practice that the repeated measurements of a quantity deliver values which are slightly different from each other. These errors do not have any fixed pattern. Since these errors occur randomly, they are known as random errors. This mainly depends on the error in the measuring process and also on the individual measuring person.

Interestingly, random errors are governed by chance; therefore, it is possible to minimise them by repeating the measurements many times and then taking the arithmetic mean of all the measurements as the correct value of the measured quantity.

If $P_1, P_2, P_3, \ldots\ldots, P_n$ be the values obtained in several measurements, then the best possible value of the quantity is given by Equation (3.1).

$$P_{\text{mean}} = \frac{P_1 + P_2 + P_3 + \ldots\ldots + P_n}{n} = \frac{1}{n}\sum_{j=1}^{n} P_j \tag{3.1}$$

This method of minimizing the random errors is based on the fact that it is rational enough to assume that individual measurements are as likely to underestimate as to overestimate the value of the physical quantity.

There are three ways of expressing an error. These are stated as below:

(I) *Absolute error* – This error of measurement is the magnitude of the difference between the value of the physical quantity and the individual measured value. Since one is not sure of the correct value of the physical quantity, the arithmetic mean P_{mean} or $\bar{P}$, is chosen to be the correct or the true value. The absolute errors in measurements are given by:

$$\Delta P_1 = P_{mean} - P_1 \tag{3.2}$$

$$\Delta P_2 = P_{mean} - P_2 \tag{3.3}$$

And so on to $\Delta P_n = P_{mean} - P_n$ (3.4)

If the arithmetic mean of all the absolute errors is taken as referred from Equation (3.2) to (3.4), then the final absolute mean ΔP_{mean} becomes as follows:

$$\Delta P_{mean} = \frac{|\Delta P_1| + |\Delta P_2| + \dots\dots\dots + |\Delta P_n|}{n} = \frac{1}{n}\sum_{j=1}^{j=n}|\Delta P_j| \tag{3.5}$$

It follows from the above-mentioned equations that any single measurement of P has to be such that Equation (3.6) is satisfied.

$$P_{mean} - \Delta P_{mean} \leq P \leq P_{mean} + \Delta P_{mean} \tag{3.6}$$

(II) *Relative Error* – It is defined as the ratio of the mean absolute error and the value of the physical quantity being measured. It is represented as Equation (3.7).

$$Relative\ Error = \frac{\Delta P_{mean}}{P_{mean}} \tag{3.7}$$

(III) *Percentage Error* – Percentage Error is the percentage representation of the relative error. It is the relative error or percentage error and not the absolute error which shows the accuracy of any measurement. This is represented by Equation (3.8).

$$Percentage\ Error = \frac{\Delta P_{mean}}{P_{mean}} \times 100 \tag{3.8}$$

Example 3.2

In a certain experiment of measuring thickness of a rod by using a micrometre, the values were found to be (in cm) 1.54, 1.53, 1.44, 1.54, 1.56 and 1.45 in successive measurements. Calculate the

(a)   mean value of thickness of the rod
(b)   absolute error in each measurement
(c)   mean absolute error
(d)   percentage error

SOLUTION

(a)   Let the thickness of the rod be

d. Then, $\bar{d} = \dfrac{1.54+1.53+1.44+1.54+1.56+1.45}{6} = \dfrac{9.06}{6} = 1.51$

(b) $1.51-1.54 = -0.03$, $1.51-1.53 = -0.02$, $1.51-1.44 = +0.07$, $1.51-1.54 = -0.03$, $1.51-1.56 = -0.05$ and $1.51-1.45 = +0.06$

(c) $\overline{\Delta d} = \dfrac{|-0.03|+|-0.02|+|+0.07|+|-0.03|+|-0.05|+|+0.06|}{6}$

$= \dfrac{0.03+0.02+0.07+0.03+0.05+0.06}{6} = 0.04$

(d) $\Delta d = \dfrac{0.04}{1.51} = 0.03$

(e)   Percentage Error $= 0.03 \times 100 = 3\%$.

Percentiles: Percentiles are the statistical values of a distribution of variables transferred unto a hundred scales [1]. The population is divided into 100 percentage categories, ranked from least to highest, with respect to some specific types of body measurements. The first percentile of any height indicates that 99% of the population would have heights of greater dimensions than that. Similarly, a 95th percentile height would indicate that only 5% of the study population would have the same or lesser heights. Similarly, 50th percentile value represents closely the average which divides the whole study population into two similar halves with one half higher and another half with lower values in relation to the average value. The percentile value is the plot of statistically treated anthropometric dimensions on a scale of 0 to 100 [2].

As an example, if a designer wants to design a seat for the Wester Commode (WC) for the common man, one needs to have complete information on the maximum and minimum values of the relevant anthropometric dimensions. The design should be such that it accommodates all categories of people, whether fat or thin, tall or short, male or female, etc. However, the choice of the percentile value for a relevant design application remains a challenge. One of the common mistakes which designers make is assuming that the 50th percentile is an apt choice for a design solution. This is incorrect. Suppose we choose to design a handle grip of any object for the common people. Now, if we choose the 50th percentile, then we will be eliminating a 50th percentage of the people trying to use the product. In this case, we will try to choose 95th percentile value or an adjusted value of more than the 50th percentile so that the design solution caters to the need of all users, as is mandated for common use by all. The reason for the mistake of opting for 50th percentile assumption as the most common case is because most of the distribution of the anthropometric data measurements fall as a part of the normal distribution (or the bell-shaped curve), where the tip of the maximum lies somewhere close to the centre point.

Further, the anthropometric data is collected separately for males and females. For some cases, the combined data is also considered wherein the maximum and minimum values are chosen from the total data set of both genders. Interestingly, while marking the values of the percentile in a table, it is a practice that we consider the 5th percentile and then the 25th percentile subsequently to the 50th percentile. This is done so because it is observed that the change in the dimensional value of the anthropometry under consideration does not change significantly between the 5th percentile and the 25th percentile. While considering the percentile values, the somatotypes of the human body is included in the tables created.

Standard Deviation: A Gaussian curve with scores in X-axis and frequency in Y-axis, where the majority of scores are near the centre of the curve and the remaining few are scattered on both the ends and follow a bell-shaped curve is not always a feasible outcome. This is because measurements of human dimensions vary significantly based on every individual, and coverage of unbiased sample sizes is also not always possible while surveying [3, 4]. In principle, the standard deviation is the measure of the variation or deviation or dispersion from the mean of the value of the data set into consideration. A low value of the standard deviation reflects that the values tend to be close to the mean (also known as the expected value) of the data set. A high value of the standard deviation indicates that the values are spread out or are dispersive in nature over a wider range.

Interestingly, another term 'Variance' is closely related to the Standard Deviation (SD). Variance is the measure of how notably a collection of data is spread out. If all the data values are identical, then it indicates the variance is zero. All non-zero variances are considered to be positive. A little variance represents that the data points are close to the mean, and to each other, whereas if the data points are highly spread out from the mean and from one another, it indicates high variance. In short, the variance is defined as the average of the squared distance from each point to the mean.

Standard deviation (denoted as ±SD) is the most precise measure of the variability of the distribution, and is presented along with the Mean value as Mean±SD. If the curve follows an ideal bell-shaped distribution, then the range of values within Mean±SD would include 68% of the population scores or the data set values with 34% on each side of the Mean. This data is based on the anthropometric data of Indian elderly collected.

For an ungrouped data source, the SD could be compiled as

$$SD = \sqrt{\frac{\Sigma(x-y)^2}{N}} \tag{3.9}$$

where

- x stands for the individual score
- y is the mean value of scores
- N is the total number of observations
- x–y is the deviation of a score from the Mean

- $(x - y)^2$, as some of the (x–y) may show a negative value, the square figures of each (x–y) are useful for computation which is finally going to be treated by the square root at the last stage of computation.
- $\Sigma(x - y)^2$, the sum of all the individual $(x - y)^2$ values

If the sample size is less than 20, then N (the total number of observations) is replaced by N−1 for getting high accuracy of SD value.

The average value of any measurement with the ±SD value in any design consideration would give a feeling of appropriate range selection for adjustability with allowances towards satisfying the required coverage of the whole population group. In case of Anthropometric Data collected, the discrete definition of the SD is considered.

This book presents a detailed study of various design elements with reference to important cases. The first case is the Re-design of the Walking stick for an average Indian Elderly. The second case is the Re-design of the Umbrella to cater to the needs of the Average Indian Elderly. Thus, the compiled data (with relevant descriptions of anthropometric terminology and illustrative descriptions with the respective measurement landmarks) are present herewith as Indian anthropometric data of elderly separately for Males, Females and Combined (males and females both considering a single study population) forms in the sequence of illustrative measurement landmarks, brief definition of dimensions/measurements and the data table.

Example 3.3

A certain sample size of an experiment to measure height of various people comprises five elements with their length measurements (in metres) as 4, 2, 5, 8 and 6. What is the Standard deviation for the experiment?

SOLUTION

$$Y_{mean} = \bar{Y} = \frac{Y_1 + Y_2 + Y_3 + Y_4 + Y_5}{5} = \frac{4 + 2 + 5 + 8 + 6}{5} = 5$$

$Y_n - \bar{Y}$ for every sample is given as follows : −

$$Y_1 - \bar{Y} = 4 - 5 = (-1)$$

$$Y_2 - \bar{Y} = 2 - 5 = (-3)$$

$$Y_3 - \bar{Y} = 5 - 5 = 0$$

$$Y_4 - \bar{Y} = 8 - 5 = 3$$

$$Y_5 - \bar{Y} = 6 - 5 = 1$$

$$\Sigma\left(Y_n - \bar{Y}\right)^2 = \left(Y_1 - \bar{Y}\right)^2 + \left(Y_2 - \bar{Y}\right)^2 + (Y_3 - \bar{Y})^2 + \left(Y_4 - \bar{Y}\right)^2 + \left(Y_5 - \bar{Y}\right)^2$$
$$= (-1)^2 + (-3)^2 + (0)^2 + (3)^2 + (1)^2$$
$$= 20$$

$$S.D. = \sqrt{\frac{\Sigma\left(Y_n - \bar{Y}\right)^2}{n-1}} = \sqrt{\frac{20}{4}} = \sqrt{5} = 2.236$$

3.1.2 ANTHROPOMETRY PARAMETERS AND DATA

Among the various anthropometry parameters, four parameters are measured from the elderly, which are used to re-design the walking stick.

1. Gluteal Furrow
2. Hand breadth without thumb, at metacarpal
3. Grip inside diameter, maximum
4. Fingertip depth

1. *Gluteal furrow:* This is the furrow, formed between the buttocks and the upper thigh muscles. The elderly were first advised to wear comfortable clothes and only with their consent, the length of the gluteal furrow was measured. This can be measured using a tape which is properly calibrated. Modern tapes can also be digital in nature. However, the multiple measurements should be done and then averaged so as to rule out the possibility of any inclined measurement leading to error. Parallax and instrument errors should be avoided, wherever possible.
2. *Hand breadth without thumb, at metacarpal:* This refers to the maximum breadth across the palm at the distal ends of the metacarpal bones (where the fingers join the palm) of the index and the little finger. This can be measured using a measuring tape or a gauged ruler or Vernier calliper.
3. *Grip inside diameter, maximum:* This is the maximum inside grip diameter, measured by sliding the hand down a graduated cone until the tips of the thumb and the middle finger remain touched to each other. The practice of cone is a standard measurement method used in the collection of Anthropometric data.
4. *Fingertip depth:* Maximum distance between the dorsal (or the back side) and the palmar surfaces (a proximal extremity (base) bearing an articular surface for the distal row of the carpal bones and additional facets towards its neighbours) of the tip of the middle finger.

Table 3.1 provides a comprehensive view of the anthropometric data collected. The table provides the data pertaining to the minimum and the maximum value of the

TABLE 3.1
Statistical Representation of Data Collected

S.No	Parameters		Min	Percentiles					Max	Mean	±SD
				5th	25th	50th	75th	95th			
1	Gluteal Furrow	Male	59.8	67.5	73	77	81	87	93	76.857	6.422
		Female	61	62.87	69	72.75	74.5	78.625	81	71.75658	4.7465
		Combined	59.8	63	71	74	78.5	85	93	74.3879	6.2193
2	Hand breadth without thumb, at metacarpal	Male	6.8	7.5	8.1	8.4	8.8	9.5	10.3	8.466667	0.6108
		Female	6.5	6.9	7.4	7.65	8.1	8.615	8.9	7.705128	0.4992
		Combined	6.5	7	7.6	8.1	8.5	9.21	10.3	8.093082	0.6762
3	Grip Inside diameter, Maximum	Male	2.5	2.8	3.5	4	4.2	5.4	5.9	3.902469	0.6967
		Female	2.1	2.5	2.9	3.3	3.7	4.3	5.2	3.318182	0.6105
		Combined	2.1	2.5	3.1	3.6	4.1	4.615	5.9	3.617722	0.7182
4	Fingertip depth	Male	1.2	1.3	1.5	1.7	1.8	2	2.1	1.661728	0.2231
		Female	0.9	1.2	1.3	1.5	1.675	2	2.1	1.515385	0.2471
		Combined	0.9	1.2	1.4	1.6	1.8	2	2.1	1.589937	0.2463

Min – Minimum; Max – Maximum; Mean – Simple Average; SD – Standard Deviation

anthropometric data, the percentile from 5th to the 95th (as is the usual practice), the Mean of the data collected and the Standard Deviation. The tabular data contains the first Standard Deviation for each data set. All the measurements are strictly in centimetre scale. This data is collected from various cities of India and specifically from elderly Indian citizens belonging to the age group 60 years of age and more. Due ethical clearance was taken before collecting the data, and the consent from every individual was taken to collect the data. A total of 159 elderly citizens participated in the measurement process with 81 males and 78 females. This is already deliberated at the beginning of the chapter itself.

In Table 3.1 Standard Deviation is calculated using Excel by using the formula STDEV.P, which is used for calculating standard deviation of the population. The details of this command can be found from EXCEL help also. STDEV.P assumes that its arguments are the entire population. If the data represents a sample of the population, then one can compute the standard deviation using STDEV. This command ignores logical values and text.

The graphs in Figures 3.1 to 3.12 are also normally potted based on the data collected to observe the mean and the Standard Deviations. This is done individually for male, female and combined for the different anthropometric parameters.

All the figures are plotted using MATLAB. It is not mandatory to use any specific tool. The figures from the data collected can be plotted in any figure plotting tool based on the convenience of the users. The figures also provide a detailed input of the second Standard Deviation so as to predict any mismatch in the data collected, which may prevent the propagation of the errors in the anthropometric data so collected the design solutions provided based on them.

The percentile is calculated from the formula as mentioned in Equation (3.10).

$$P_p = I + \frac{PN - F}{f_p} \times i \tag{3.10}$$

where

P_p = required percentile rank.
I = lowest value of the class interval where P_p falls.
PN = cumulative frequency in relation to the P_p point
F = cumulative frequency below the lowest value of the P_p class
f_P = actual frequency of the P_p class
i = class interval score

It is interesting to note that shorter the class interval, the more accurate the results are received or calculated. This is as referred from [1].

Gluteal furrow male: For the Gluteal furrow for the males, the calculated details are as follows:

Mean = 76.857, SD = 6.422, Mean − SD = 70.435, Mean + SD = 83.279, Mean − 2*SD = 64.013, Mean + 2*SD = 89.701. All the data shown in Figure 3.1 are in centimetres.

GLUTEAL FURROW MALE

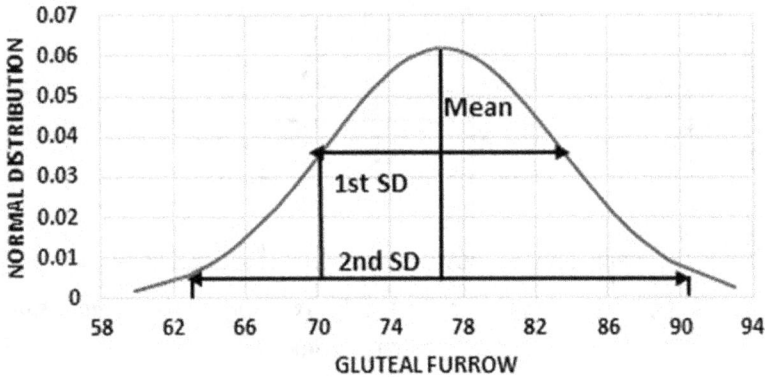

FIGURE 3.1   Graph of gluteal furrow male data using mean and standard deviation (SD).

Gluteal furrow female: The Gluteal furrow for the females and their calculated details are as follows:

Mean = 71.756, SD = 4.746, Mean − SD = 67.01, Mean + SD = 76.502, Mean − 2*SD = 62.264, Mean + 2*SD = 81.248. All the data shown in Figure 3.2 are in centimetres.

Gluteal furrow combined: The Gluteal furrow for the males and females can also be taken together and combined. Their calculated details are as follows:

Mean = 74.387, SD = 6.219, Mean − SD = 68.168, Mean + SD = 80.607, Mean − 2*SD = 61.949, Mean + 2*SD = 86.826. This is based on the plot in Figure 3.3. Interestingly, the Gluteal furrow of the males is more than that of females, which is also in tandem with the recorded observations

GLUTEAL FURROW FEMALE

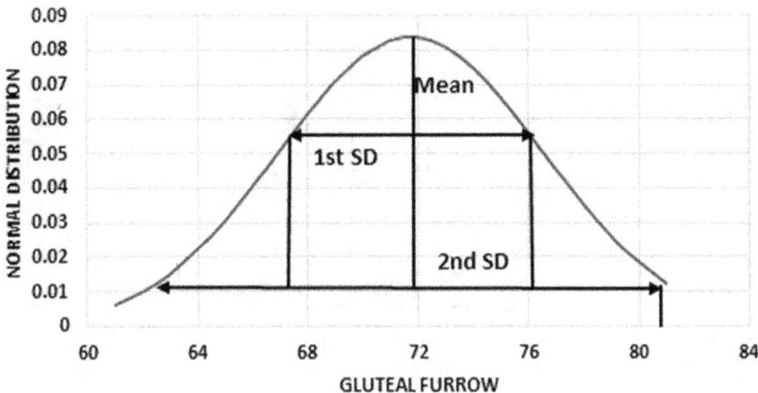

FIGURE 3.2   Graph of gluteal furrow female data using mean and standard deviation (SD).

GLUTEAL FURROW COMBINED

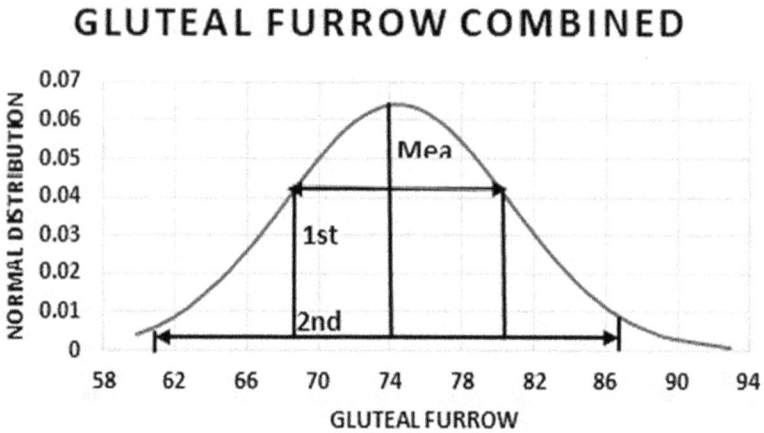

FIGURE 3.3   Graph of gluteal furrow of the combined data of both male and female taken together using mean and standard deviation (SD).

for general Indian anthropometry as recorded in the book [1]. Further, the graphs follow the bell-shaped curve showing that the recorded data follow normal distribution.

Hand breadth male: Based on the data collected from the field study, the calculated details are as follows:

Mean = 8.466, SD = 0.6108, Mean – SD = 7.855, Mean + SD = 9.076, Mean – 2*SD = 7.244, Mean + 2*SD = 9.687. All the details of this data are referred from Figure 3.4. Since the variations of the hand breadth do not change very significantly unlike the gluteal furrow, the normal distribution curve is narrower as compared to the previous case plots.

HAND BREADTH MALE

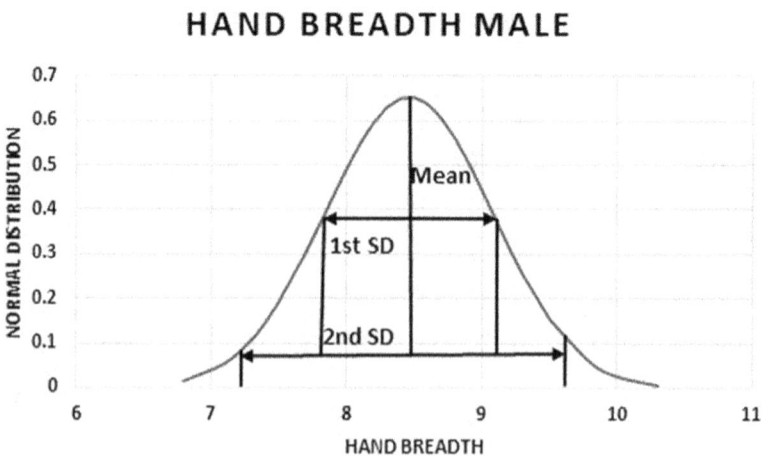

FIGURE 3.4   Graph of hand breadth male data using mean and standard deviation (SD).

Hand breadth female: Based on the data collected from the field study, the calculated details are as follows:

Mean = 7.705, SD = 0.4992, Mean − SD = 7.205, Mean + SD = 8.204, Mean − 2*SD = 6.706, Mean + 2*SD = 8.703. The above details are referred from Figure 3.5.

Hand breadth combined: The combined data of both males and females taken together is as shown in Figure 3.6. The handbreadth measurements of the female elderly are found to be less than the male handbreadth measurements. The various calculated parameters based on the measurements are as follows:

HAND BREADTH FEMALE

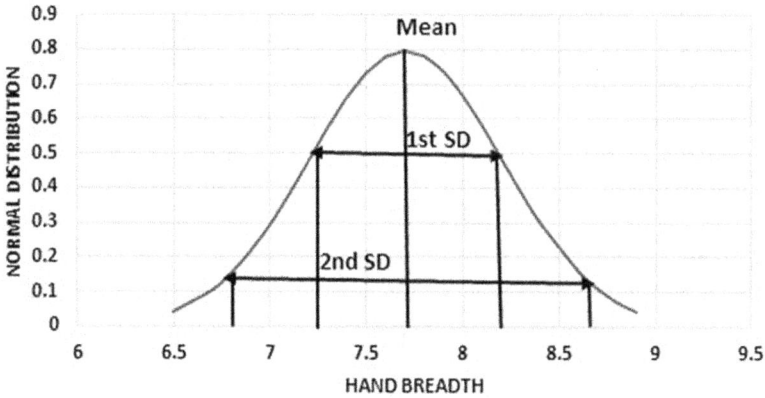

FIGURE 3.5 Graph of hand breadth female data using mean and standard deviation (SD).

HAND BREADTH COMBINED

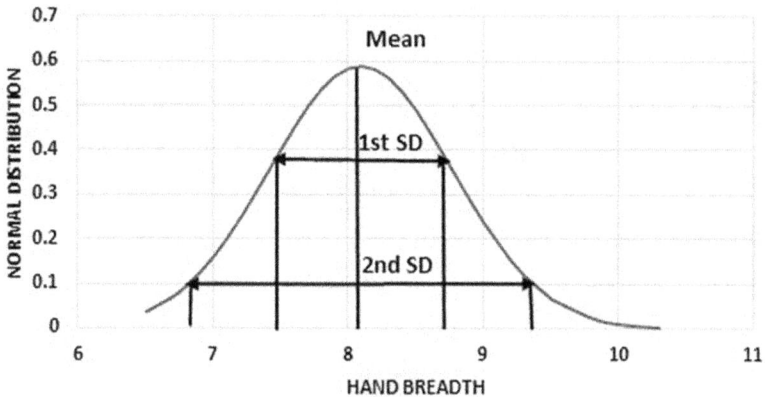

FIGURE 3.6 Graph of hand breadth combined data using mean and standard deviation (SD).

Mean = 8.093, SD = 0.6762, Mean – SD = 7.416, Mean + SD = 8.7692, Mean – 2*SD = 6.740, Mean + 2*SD = 9.445

Grip diameter male: The field study of the Grip diameter of Male elderly participants/users reveals the following calculated parameters:

Mean = 3.902, SD = 0.6967, Mean – SD = 3.205, Mean + SD = 4.598, Mean – 2*SD = 2.508, Mean + 2*SD = 5.295. The data of the plot is shown in Figure 3.7.

Grip diameter female: The field study of the Grip diameter of Female elderly participants/users reveals the following calculated parameters:

Mean = 3.318, SD = 0.6105, Mean – SD = 2.707, Mean + SD = 3.928, Mean – 2*SD = 2.096, Mean + 2*SD = 4.539. The data of the plot is shown in Figure 3.8.

GRIP DIAMETER MALE

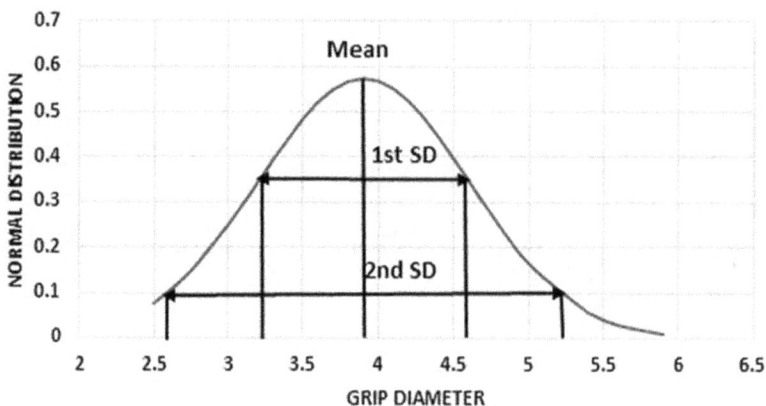

FIGURE 3.7   Graph of grip diameter male data using mean and standard deviation (SD).

GRIP DIAMETER FEMALE

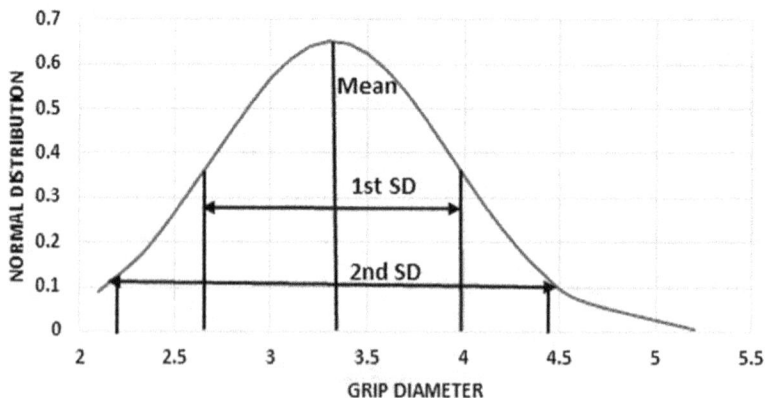

FIGURE 3.8   Graph of grip diameter female data using mean and standard deviation (SD).

Grip diameter combined: The combined data of both males and females taken together is shown in Figure 3.9. The grip diameter measurements of the female elderly are found to be less than the male grip diameter measurements. The various calculated parameters based on the measurements are as follows:
Mean = 3.6177, SD = 0.7182, Mean − SD = 2.898, Mean + SD = 4.335, Mean − 2*SD = 2.281, Mean + 2*SD = 5.053

Fingertip depth male: Based on the measurements carried out for Fingertip depth in males, reveals the following statistical data:
Mean = 1.661, SD = 0.223, Mean − SD = 1.438, Mean + SD = 1.884, Mean − 2*SD = 1.215, Mean + 2*SD = 2.107. The fingertip-measured data details are plotted in Figure 3.10 respectively.

GRIP DIAMETER COMBINED

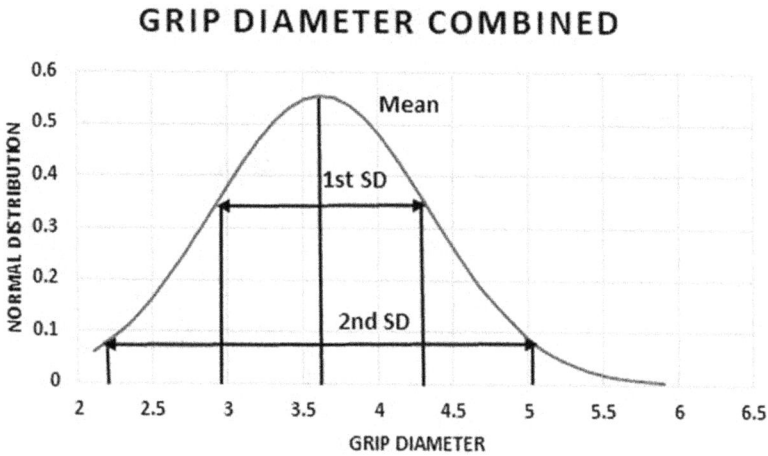

FIGURE 3.9   Graph of grip diameter combined data using mean and standard deviation (SD).

FINGERTIP DEPTH MALE

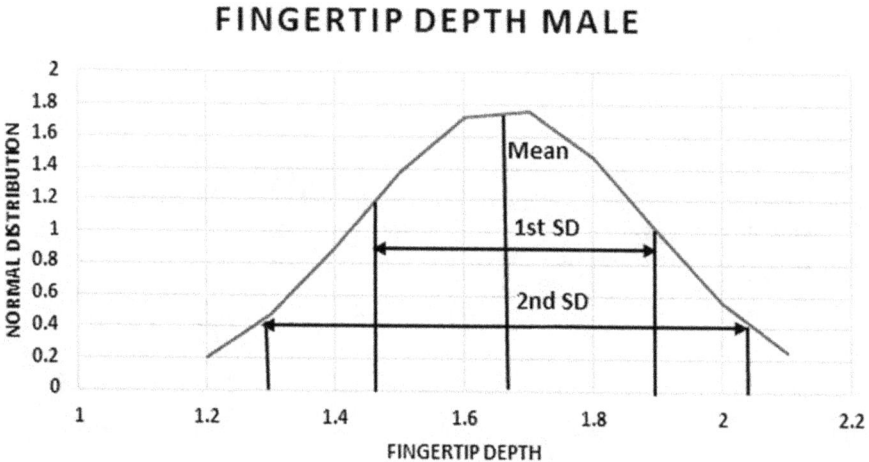

FIGURE 3.10   Graph of fingertip depth male data using mean and standard deviation (SD).

Fingertip depth female: Following the same protocol, the statistical data of the fingertip measurements in females reveal the following details:

Mean = 1.515, SD = 0.247, Mean − SD = 1.268, Mean + SD = 1.762, Mean − 2*SD = 1.021, Mean + 2*SD = 2.009. This can be seen in the plot of Figure 3.11.

Fingertip depth combined: The combined data of elderly males and females taken together is statistically treated and plotted. The study shows the following details:

Mean = 1.589, SD = 0.246, Mean − SD = 1.343, Mean + SD = 1.835, Mean − 2*SD = 1.097, Mean + 2*SD = 2.081. This is also revealed from the plot of Figure 3.12.

FINGERTIP DEPTH FEMALE

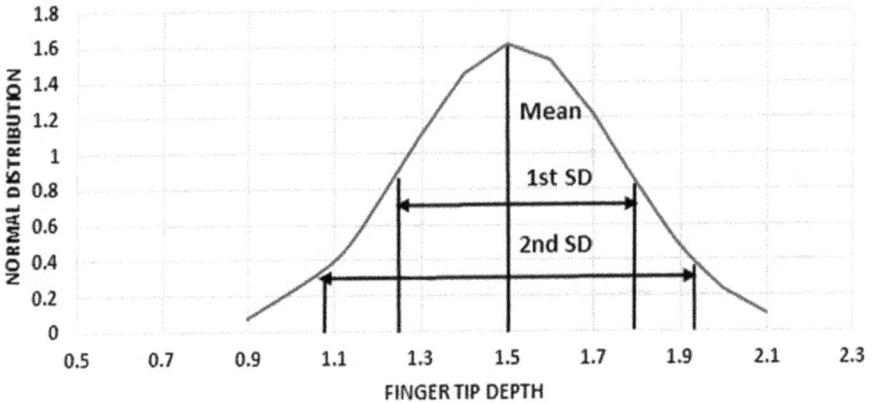

FIGURE 3.11   Graph of fingertip depth female data using mean and standard deviation (SD).

FINGERTIP DEPTH COMBINED

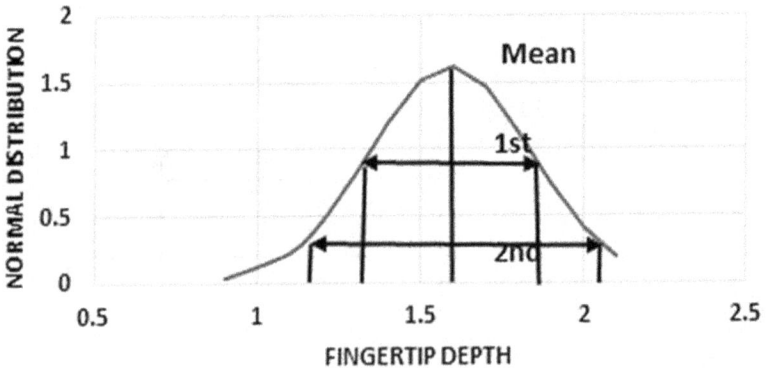

FIGURE 3.12   Graph of fingertip depth combined data using mean and standard deviation (SD).

If a data distribution is approximately normal, then about 68% of the data values are within one *standard deviation* of the mean (mathematically, Mean±SD, where mean is referred as the arithmetic mean), about 95% are within two *standard deviations* (Mean±2*SD), about 99.7% are within three standard deviations (Mean±3*SD). The graphs consist of only one and two standard deviations of all anthropometry references. Interestingly, on investigating the graphs, it is found that all these observations are valid for the user data measurements collected from the field study of the Indian elderly. The anthropometric data collected is extremely crucial for the design solutions that shall be rendered after the Direct Observation and Analysis is carried out to find the responses by the Indian elderly on the uses of the walking stick in their lives.

3.2 DESIGN OF THE QUESTIONNAIRE FOR DIRECT OBSERVATION AND ANALYSIS

Direct Observation and Activity Analysis have been used to conduct the initial user study for seeking the limitations and advantages of the existing walking sticks and umbrellas being commercially used. The elderly were given a choice between different walking sticks, their body discomfort mapping on rigorous use along the palm, hands and shoulders were mapped.

The total sample size of initially 20 was chosen for rigorous and extensive users of the walking stick among the elderly. It is observed that the responses received by the 20 participants almost saturate after some questions and so 20 was an optimum data set for the Interview based on the questionnaire framed.

An important context of the users of the walking stick was the users are generic users of the walking stick who use the product due to age-related issues. Also, the opinion and inputs from the geriatrics who have undergone any medical emergency like knee replacement or bone grafting or surgeries related to the lower part of the body have also been taken so as to add modularity to the walking stick in the Indian context.

It was interesting to note that saturation in the replies was reached after 15 users, which is then used for brainstorming in the design process. The questionnaire designed for interviews included both qualitative and quantitative questions. Certain qualitative questions for seeking opinion on a Likert scale of 7 (as an average) have also been chosen for walking stick uses [5–7]. The questions along with relevant explanations are clearly marked with each question. The design of the questionnaire has been done with an aim to pursue the critical inputs from the elderly users about their connectivity with the product as well.

The qualitative questions are highlighted in bold and italics. The quantitative questions are highlighted in bold and underlined as well. There are some notes added shown in bold plus italic plus underlined. The notes describe the need of the question or the context under which the question is relevant for the design study. This helps in understanding the user better. However, in the whole process of interview using the questionnaire, the age of the user is asked towards the end so as to encompass the effects of the Hawthorne effect. Hawthorne effect is the alteration of the behaviour of the user or subjects due to awareness of being observed [8, 9]. This may lead to inaccurate recording of responses.

The relevant figures used with each question are shown in Figures 3.13 to 3.15. These figures were also shown to the users in the field study during the interview session to seek deterministic inputs from them for modifying the walking stick or to locate/co-locate the pain points from the users while using the walking stick.

The detailed questionnaire along with the reasons for posing the question is given below:

Age:......................... (It is a Quantitative question which is asked towards the end of the interview session to avoid the Hawthorne effect. If this question is posed at the beginning of the interview itself the likelihood that the responses received by the elderly shall be biased is likely to be very high.)

Gender:..................... (It is a Quantitative question which is marked or used for data analytics and ideation after the interview session of all the interviewees is completed. Sometimes the preferences of the male and female choices differ based on Gender and so it constitutes an important point in the questionnaire.)

1. How long have you been using the walking stick?? (It is a Quantitative Question to be plotted on the Bar chart to enable the designer to know of the depth of need of the walking stick. This also gives information on whether the elderly is totally dependent on it or uses it only occasionally as and when the requirement is.)

2. What was the reason for using a walking stick? *(This is a Qualitative Question where each response is noted. This question helps identify the context of use of the walking stick by the elderly.)*

3. Which among the given is the handle of your stick? (This is a Quantitative question which is to be plotted on the Bar Graph. Figure 3.13 shows the different types of commonly used handles of the walking sticks. The elderly is shown the image of the handle and then asked to make a choice or preference also. Figure 3.13a refers to the conventional U-shaped handle, which is usually present in the 'lathi' or the wooden stick design. Figure 3.13b is an L-shaped handle having a circular curvature for an improved grip. This is available with both standard wooden sticks and

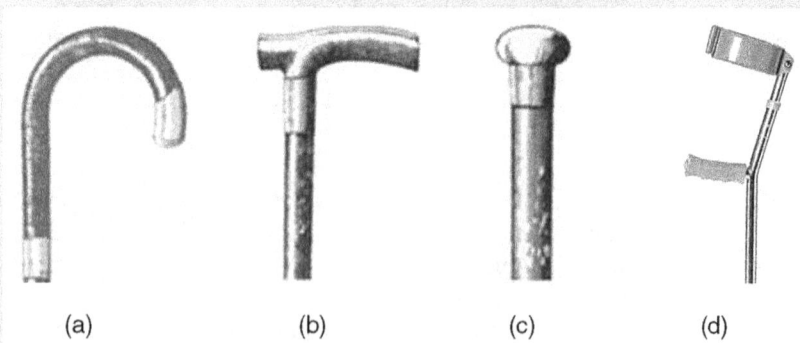

(a)   (b)   (c)   (d)

FIGURE 3.13   Different types of handles of walking sticks.

also with aluminium-based adjustable walking sticks. Figure 3.13c has a round or spherical top with no extended handle. Usually these kinds of handles are preferred to show a certain level of dominance or aristocracy. This kind of handle is used for short-term walking support. Figure 3.14d is inspired from the crutches and is an extended version of the handle. This is used usually in places where the mobility and the body weight of the user are not to be placed on the limbs at all due to some medical exigency condition. Most people with some permanent trouble of walking are suggested to use walking sticks with such handles.)

4. Which among the given is the grip of your stick? (This is a Quantitative question which needs to be plotted as Bar Graph. Figure 3.14 shows the various bases which are available with the various walking sticks. The comfort level in using them is also asked and the choice or preference of usage of the base is also asked to the elderly. The elderly is shown the picture of the various bases to make a choice among them which they are currently using or prefer to use in future. Figure 3.14a shows a conventional round base which is usually present with wooden sticks. The base cane is made of rubber in order to improve the grip between the walking stick used as a support and the ground plane, increasing the frictional contact. Figure 3.14b shows an extended version of the conventional round base with L-shaped slope so as to increase the surface area of contact between the walking stick base and the ground. Figure 3.14c shows a triangular base. Interestingly the footprint of the base area is increased, which helps to improve the support system using the walking stick. A further modification of the walking stick base is by adding another leg to the base and making it four-legged base. This provides a firm base, and the walking stick can be left standalone without any chance of a fall if left out ideal. Such a walking stick is preferred by people who had surgery and have to walk after the surgery to restore the normal lifestyle.)

5. How many times do you use the stick for walking in a day? (This is a Quantitative question and is to be plotted on the Bar Graph. This also gives information of whether the elderly is totally dependent on it or uses it only occasionally as and when the requirement is.)

(a) 1

(b) 2–5

(c) 6–10

(d) Every time

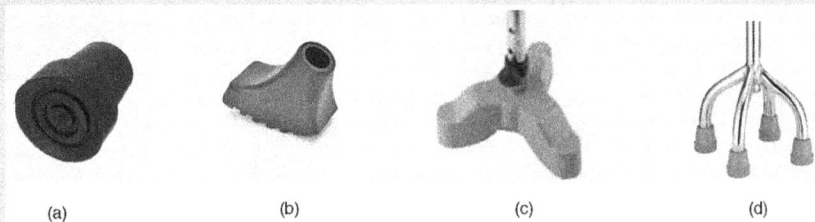

(a)   (b)   (c)   (d)

FIGURE 3.14 Different types of base grips of walking sticks.

6. Mark the area where you feel pain/discomfort while holding. (This is a Qualitative Question where user marks the area of discomfort)
   (This is called Body Discomfort mapping. Further, all four options to explore Left handed and Right-handed users are incorporated.)

7. On a scale of 1–7, how much pain/discomfort do you feel?
   (Least) 1-----|-----|-----|-----|-----|----- 7 (Most)
   Likert Scale: Quantitative analysis of the Qualitative data, user maps the rate of discomfort. Scale of 1–5 is not chosen since it is not clear that some people might have injuries and so it might lead to inherent pain which cannot be mapped on a scale of 1–5. Scale of 1–10 is not chosen since this scale is chosen for more complex scales and since this is a focused group on elderly, scale of 1–10 is avoided.

8. Do you use the stick outside the house?
   (a) Yes
   (b) No
   (This is a Quantitative question which can be plotted on the Bar Graph)

9. Where do you store the stick while…? (This is a Qualitative question to aid in designing some compact design for the Walking Stick which is easily portable also.)
   (a) at home: _____
   (b) traveling: _____

10. Do you have a problem finding the stick at home?
    (a) Yes
    (b) No
    (This is a Quantitative question that is to be plotted on the Bar Graph. This question also aids to find design solution for feedback to locate the product.)

11. What problems do you face while walking with the stick? *(As a Qualitative question, the user inputs are to be mapped for proposing an efficient design.)*

12. What do you feel about your walking stick? *(This is a Qualitative question that provides the emotional connect of the elderly with the product. This is extremely critical in any user-centric design or design thinking process.)*

13. How can the walking stick be improved? *(This is a Qualitative question where preliminary suggestions are sought from the elderly users themselves)*

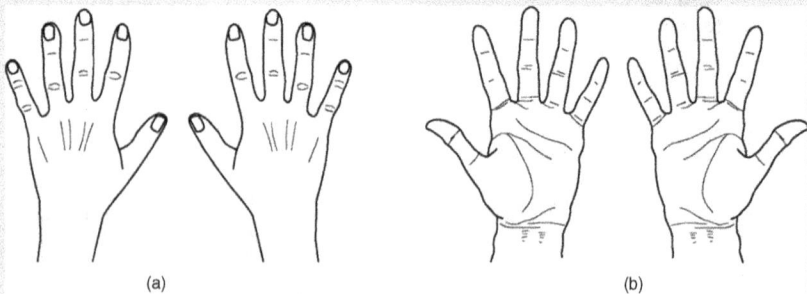

FIGURE 3.15   Pain point mapping on the hand (a) Dorsal side of hand (b) Palm side of hand.

The Likert scale of 1–7 is also discussed with its reason of choice of such a scale. The questionnaire is a part of the Direct Observation and Activity Analysis for understanding the various usage of the Walking stick by the elderly.

3.3 DESIGN THINKING APPROACH USING DOUBLE DIAMOND AND EMPATHY MAP

After the Anthropometry data is available and the Questionnaire is designed, the double diamond map approach of Design Thinking is used [10–13]. This approach for Walking Stick design is also adopted in [14]. The design thinking model and the concepts are also presented in [14]. The method is already explained in Chapter 1.

The various elements of the Double Diamond model are as mentioned below:

1. *Discover (Research):* The design process according to the double diamond model begins with the discovery of user needs, current design and choice of various methods for the same. The popularly used walking sticks are presented in [15–20]. These are broadly classified as below: -

   Illuminating walking stick: This consists of a tubular cane shaft, a handle having an angled section extending laterally from one end of said cane shaft, a first light source disposed within said cane shaft adjacent to said translucent section for emitting flashes of light, a battery housed within the cane for supplying power to the already present first and second light source.

   Pick-up type walking stick: A walking stick which is hollow and is provided with pick-up mechanism consisting of fingers which separate apart when projecting from the lower end of the stick but close together when retracted into the stick so as to be effective when picking up articles from the ground.

   Walking stick: A simple walking stick consists of an elongated support element comprising an elongated hollow stick having upper and lower parts which are provided on its lower part with two supporting legs which can be tilted or moved angularly between a folded-in position and a folded-out, active position.

   Multifunctional walking stick: This walking stick can have multiple functionalities. The focus of its electric torch can be adjusted by a finger at any time for pinpointing a small location or for shining a wide area. The stem is adjustable by double-safe construction for keeping adjusted stem firmly in position and is more solid and stable than any prior art after adjusting.

The above are some of the unique walking stick solutions which are available for commercial or research use. Based on the available literature survey and data, the methodology followed is the same as described in Chapter 1.

With reference to the pain points of usage of walking stick, the methodology followed is as below:

Shadowing: To implement shadowing, the elderly were not asked their names and age until the end of interview so that they do not become aware as a part of the Hawthorne effect. Had this information been sought at the outset of the interview itself, the users would have been made conscious or aware leading to erroneous information instead of honest feedback or reply.

Interviews: Semi-structured interviews were conducted with all the participants (in this case, the male and female elderly) of the study to collect qualitative and quantitative data among different categories: objective measures of the environment, resident perceptions and attitudes about the environment and walking stick, walking behaviour data, pain reception and habit formation. In a lot of the cases, the interview was the correct alternative to the questionnaire and was narrated. The narratives were mostly in a discussion mode so as to make the participants comfortable enough for seeking honest opinion or feedback.

Questionnaires: A set of questions was designed beforehand for obtaining statistically useful or personal information from individuals. The questionnaire had both qualitative data and quantitative data for collecting valuable information.

Qualitative data and quantitative data were mapped to understand the patterns of usage, pain and the physical or cognitive activity involved.

Likert scale: Qualitative data describing the pain felt on various parts of the hand were also quantified using the Likert scale. It is a psychometric scale commonly involved in research that employs questionnaires. It is the most widely used approach to scaling responses in survey research. In the proposed study, no pain was mapped as being zero or pain not felt at all, and ten was considered for being extremely painful.

The various pain points were then marked on the image of the palm by the users to describe the regions where they felt it. This is clearly shown in Figure 3.16. Some of the areas are overlapped. Superposition of the areas identified by the participants yields those sections in the palm where maximum pain is felt.

2. *Define (Insights):* This refers to defining the areas of interest and development by analysis of data collected in the previous stage. Factors which are considered in this stage include Storage, Reason for use, Parallel tasks, Illumination and the Terrain. Based on the interviews conducted, the various factors to be attributed to design issues are stated below:

Storage: Storage of the stick for the primary user of the stick in different contexts of use and settings is taken into account. This also gives an

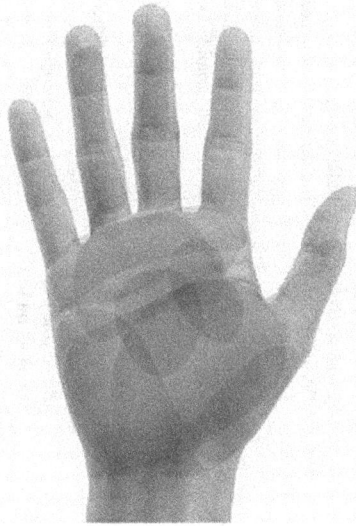

FIGURE 3.16 Various locations reported of pain while using the walking sticks.

insight into the varying settings where the aid can be used and how its use changes from one setting to the other.

Reason for use: The most common reasons for use include, but are not exclusive to, maintenance of balance, confidence in walking, support of weight, prescription by doctors and ease in climbing stairs. These reasons have been identified based on the responses received during the interview sessions.

Parallel tasks: This gives an insight into how the user behaviour alters from usual, when they are given a walking stick and helps understand the extent of impact it can cause.

Illumination: In order to improve the illumination or to make the walking stick visible, the conventional stick can be white-finished so that it can be readily seen by others in the vicinity. However, it is only effective in the daytime and may be inconspicuous or invisible at night to threaten the security of the elderly person in walking.

Terrain: Residents' perceptions of their neighbourhood environment and their physical ability have a direct influence on their walking behaviour based on the terrain. This is also been clearly identified in the responses recorded during the interview sessions.

The consolidated response of all the 20 users and the elderly interviewed is tabulated in Table 3.2. The type of handle and the type of base grip is referred from Figure 3.13 and Figure 3.14, respectively, from the questionnaire designed.

3. *Analysis:* Based on the inputs that are collected and tabulated, the analysis of each and every response is done. The responses are extensively

TABLE 3.2
User Study of Walking Sticks

S.No.	Gender	Age	Experience in Usage	Reason of Usage	Type of Handle	Type of Base Grip	Storage	Utility
1	Female	76	2 months	Insect bite in foot	b	c	Under the bed	Balance and confidence
2	Male	72	7–8 years	Skin disease	c	a	Under the bed	Balance and confidence
3	Female	75	9 months	Kneecap damaged	b	a	Beside the bed	Balance and confidence
4	Male	82	3–4 years	Pain in the knees	c	a	Under the bed	Balance and confidence
5	Male	70	18 years	Leg operation	c	a	Under the bed	Balance and confidence
6	Female	100	30 years	Knee joint pain	a	a	Under the bed	Balance and confidence
7	Female	73	2 years	Knee/back pain	b	a	Under the bed	Balance and confidence
8	Female	82	2 years	Back bent	c	a	Under the bed	Posture
9	Female	84	3 years	Fracture	b	a	Beside the bed	Balance and confidence
10	Male	88	15 years	Leg operation	b	a	Cup board	Support
11	Female	70	2 years	Knee injury	b	d	Under the bed	Balance and confidence
12	Female	80	1 year	Bent back	b	d	Under the bed	Posture
13	Female	72	3 years	Knee injury	b	a	Beside the bed	Balance and confidence
14	Female	85	1 year	Gap in knee	a	a	Hanging on wall	Support
15	Female	72	2 years	Knee replacement	b	a	Beside the bed	Balance and confidence
16	Female	83	11 years	Arthritis	b	d	Beside the bed	Balance and confidence
17	Female	85	4 years	Arthritis	b	d	Beside the bed	Balance and confidence
18	Female	68	8 years	Arthritis	a	a	Beside the bed	Balance and confidence
19	Female	72	2 weeks	Fracture	b	d	Beside the bed	Balance and confidence
20	Female	68	5 days	Knee operation	b	d	Beside the bed	Balance and confidence

brainstormed and detailed to avoid missing any critical information while providing design solutions. Among the first step for the analysis is mapping the pain points while using their usual walking sticks.

Users were asked about various parts of their palm which pained while using their walking stick; users then marked various spots on the image of their palm. These images were then traced and converted into 10% opacity overlays and then all marked on one common hand frame to pinpoint areas which were common in the complaint about pain. The higher the opacity of the area, the usual the spot was for pain or discomfort. This graphically explained and analysed the common pain points with similar handles and is shown in Figure 3.16.

Once the pain points are mapped, the chunking of the data is done. This is achieved through graphical chunking of the insights of the user data collected from interviews. This step included identifying the quantum of users that fall under different categories depending on their usage and preference.

The users were asked what was the reason they started using a walking stick in their daily life and what purpose did they think it served for them. Further, what they did with the stick when not in use for the short and long duration of time (minutes while sitting to hours at night). A repetition was observed in the responses after interviewing less than 25% of the participants, and thus the data were separated into major categories and plotted on a pie chart to show the similarity in user preference and behaviour.

From the plotted data, it was observed that bedside, as referred from Figure 3.17, and under the bed, are the most preferred location of storage for longer durations and that most people believe that their walking stick helps them keep balanced and gives them confidence, as shown in Figure 3.18. Figure 3.19 shows the most popular base grip available in the market, which is single base as referred from Figure 3.14, and Figure 3.20 shows the most usable handle grip, which is a standard L-shaped handle as referred from Figure 3.13.

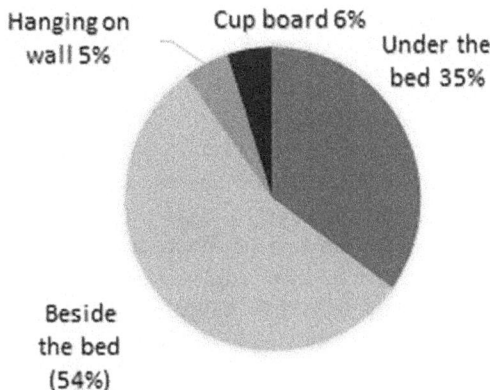

FIGURE 3.17 Graphical data of storage.

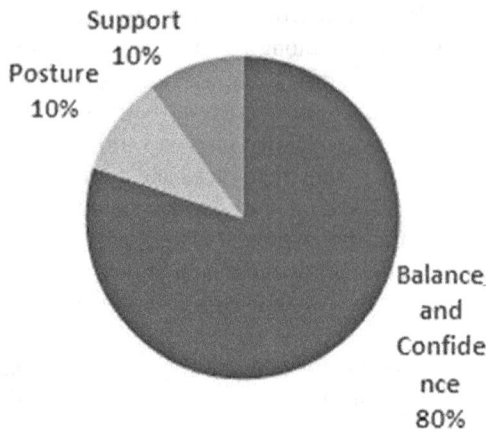

FIGURE 3.18 Graphical data of utility.

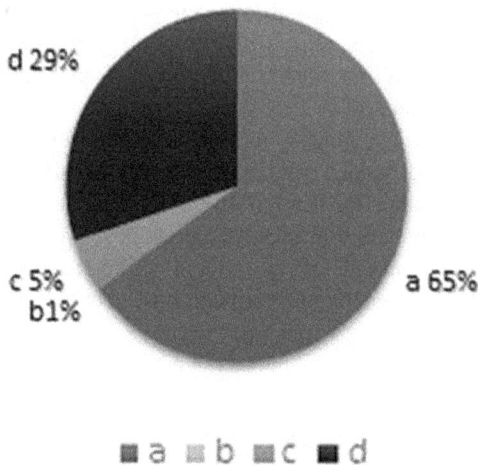

■ a ■ b ■ c ■ d

FIGURE 3.19 Graphical data of base grips (Refer Figure 3.14).

Based on the inputs received, some generic or common answers for qualitative questions are also observed. In most cases, the reason for usage is recommended by a doctor to prevent chances of a fall or to strengthen gait, that is, a person's manner of walking and thus provide confidence and independence in walking, etc. However, there are some people who despite the need of a walking stick completely avoid using it as it makes them feel overdependent and medically unfit. Also, bulky sizes and lack of some required features are among some of the reasons to avoid using them regularly. This is also in tandem with some of the reasons cited in [21, 22].

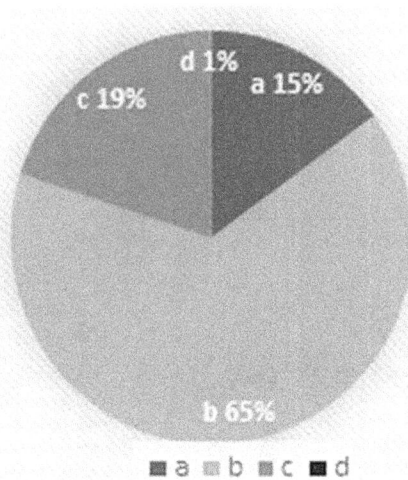

FIGURE 3.20 Graphical data of handle (Refer Figure 3.13).

4. *Empathy maps:* After the chunking of the data is done, the information is further placed in the empathy map to understand the main requirements and modifications that need to be done while re-designing the walking stick. The empathy map for the 20 users collectively is shown in Figure 3.21.

The information chunked in the various quadrants is explained below:
Says
- Says quadrant contains what the users say loud in an interview or user study.
- Here in the above empathy map, when user was asked about reason for usage, user disclosed that it was on the recommendation of the doctor because of knee operation, insect bite, fracture, back bent, arthritis and some medical problems. Age-related issues were of equal concern by the users to use the walking stick.

Thinks
- Thinks quadrant captures what the user is thinking throughout the user experience.
- Based on the observations of the users, conclusions of thinks quadrant are drawn.
- While using the walking stick,
  1. Some users think that stick might provide balance, support and confidence to them.
  2. Some users think that stick might reduce the fear of falling.
  3. Some users think that stick might reduce the pain of their limbs
  4. Some users think that stick might provide them good posture.

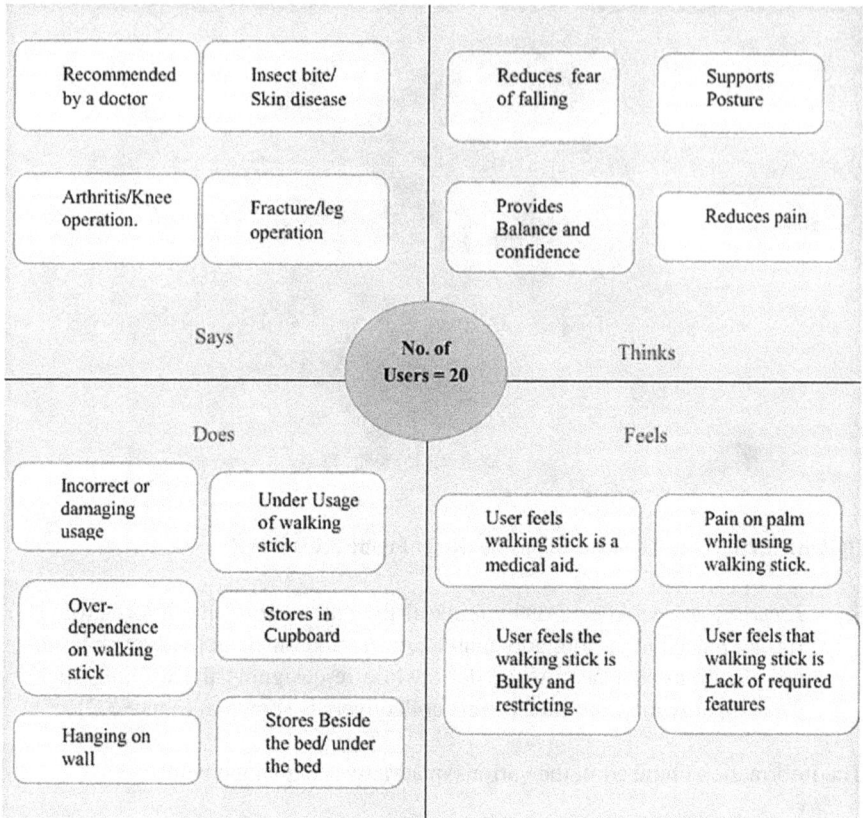

FIGURE 3.21   Empathy map of the 20 users studied in direct observation and activity analysis.

Does
- The Does quadrant encloses the actions the user takes.
- From the user study, some conclusions are drawn for the Does quadrant
- Some users do not use the walking stick properly, that is, in a way it should be used.
- Some users use their sticks incorrectly, thus damaging the walking stick.
- Some users are overly dependent on the walking stick.
- For storing or keeping or resting the walking stick after usage,
    1. Some users store it beside the bed.
    2. Some users store it under the bed.
    3. Some users hang it on the wall by the handle side, if the handle is U-shaped or bent.
    4. Some users store in the cupboard.

Feels
- The Feels quadrant connects the product, that is, the walking stick to the user's emotional state and its usage practices.

- While using the walking stick
    1. Some users feel it as a medical aid. They have stigma associated with walking stick since they feel that other people might perceive them to be seriously ill, which is why they use a walking stick. People might perceive the persona of the elderly user with walking stick to be totally dependent on a third person rather than judging them as being independents who want to avoid dependence on any third person for a walk or any other associated activity.
    2. Some users feel pain on palm when the stick is used for a long time.
    3. Some users feel the stick is bulky in size and, thus, it restricts their movement and freedom.
    4. Some users feel that their walking stick lacks various features for comfortability and emergency.

Based on all the quadrants of empathy map, the ideation process is initiated. The develop stage of the double diamond map reflects the stage where all ideations are presented. This is explained in the next section, that is, Section 3.4 on Design Concepts.

Further, deliver section of the double diamond map refers to the fabricated prototypes of the ideated concepts. The various ideations are then fabricated into prototypes which are then tested in a limited environment like the scale or for field testing, that is, the practical case. This is discussed in detail in Chapter 4.

3.4 DESIGN CONCEPTS

The Ideation process is depicted in Figure 1.6 as the 'develop' stage of the double diamond map. Various designs are worked with different feasible solutions, following the Indian anthropometric dimensions. However, certain constraints are kept in the back of the mind, which are stated as follows:

1. *Great illumination* – This is to incorporate the fact that walking stick is visible from a distance when the elderly is using it. Also, efforts to incorporate easily deployable lighting solution can be provided in the walking stick so as to facilitate walking in dark areas or especially during monsoons when power supply cut makes it difficult for the elderly to walk.
2. *Ergonomic* – The proposed walking stick design solution should be an ergonomic retrofit for elderly users.
3. *Easy storage* – One of the cognitive load on the elderly is to rest or store the walking stick after its use. This can be during the night before going to bed or while taking a short break while walking. This should be addressed in the proposed design solution.
4. *More of a fashion accessory than a medical aid* – As mentioned earlier, some elderly do not want to use the walking stick as it projects the elderly as under some medical condition. People might tend to assess them as dependents. This should be avoided in the proposed design solutions.
5. *Adjustable* – The height of an individual always differs from individual to individual. Thus, the walking stick should be adjustable to facilitate its use by different people.

6. *In tandem with the Indian elderly's mental model* – It has been presented before also that the Indian elderly tend to get attached to the products they use in day-to-day life. It is essential that the mental model about perception by the Indian elderly be incorporated in the proposed design solutions.

Based on the above inputs and the discussions, using design thinking approach, three different concepts are proposed. The various details associated with each of the design solution is also presented. The three different proposed solutions are as follows:

3.4.1 CONCEPT 1

The first proposed design keeps the simplistic approach in mind. This design is proposed with a tri-junction grip and a footrest using a conventional base of the walking stick. The proposed design is illustrated in Figure 3.22.

The key features of the proposed ideation are given as follows:

A. Textured rubber to increase tactile feedback.
B. Better grip on the stepper.
C. The concept tries to combat the stigma of associating walking aids with old age.
D. A three-pronged handle design makes it more comfortable to use – the wider spread reduces the amount of stress placed on the hand's ligaments. The design of handle is in such a way that there is a 120-degree angle between each prong.

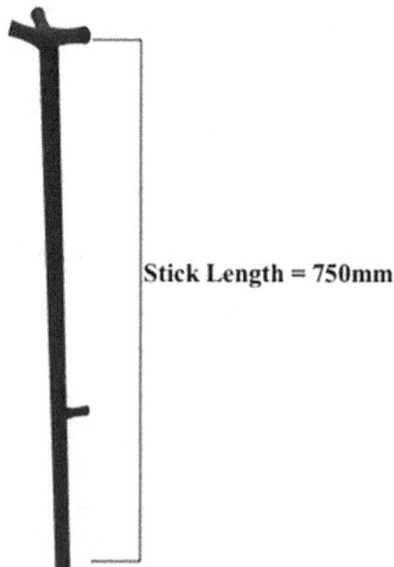

Stick Length = 750mm

FIGURE 3.22   Walking stick concept 1.

E. A three-pronged handle allows users to hook the stick onto a surface when they aren't using it, so they don't have to worry about it falling over and having to bend down to pick it back up.

F. An extruded part towards the base is used as a footrest whenever the user feels the need to rest his leg. It will not cause any interference to the user while walking and doubles as a support to hang the stick when not in use.

G. The length of the walking stick is proposed around 750 mm looking at the data collected from the gluteal furrow, so as to maximise the number of elderly who can use it comfortably.

3.4.2 CONCEPT 2

The second proposed design encapsulates the need for technology intervention by providing space for adding electronic circuitry and providing balance during walk. A footrest is also provided which makes the walking stick capable of standing on its own when left alone, so that it may not be rested against a support like a wall or bedside, etc. The proposed design is illustrated in Figure 3.23. The figure also shows various proposed dimensions based on the anthropometric data collected. The proposed dimensions may be altered by looking at the accessibility of the material to fabricate the walking stick.

The anthropometric reference for design concept 2 is tabulated in Table 3.3.

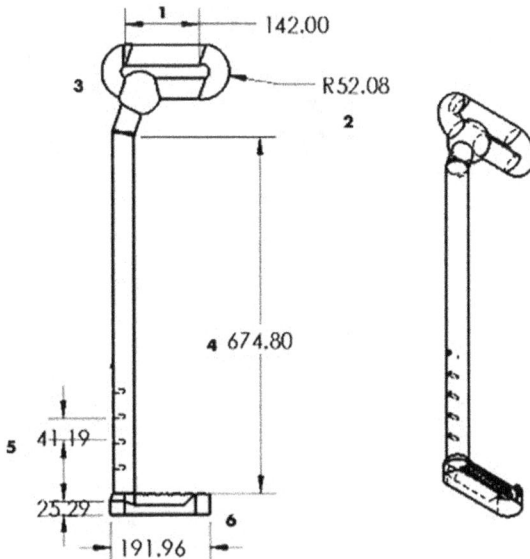

FIGURE 3.23 Walking stick concept 2 and its dimensions.

TABLE 3.3
Anthropometry Reference to Figure 3.23

Figure 3.23	Anthropometry	Dimension
Handle width	Handbreadth without the thumb, at the metacarpal	117 mm
Grip diameter	Grip inside diameter	42 mm
Cavity	fingertip depth	18 mm
Height Min	Gluteal furrow min	682 mm
Height Max	Gluteal furrow max	923 mm

3.4.3 CONCEPT 3

The walking stick concept 3 is ideated keeping in mind the fact that many elderly users do not want to carry a walking stick so as to reflect an image of an unwell or ill person. Instead of carrying it as a medical advice, if it adds to aesthetic features, then the persona of a person reflects a healthier state of mind and independent set of individuals. This concept is illustrated in Figure 3.24.

The various features of the proposed ideated concept are given below:

A. Two gripping postures are enabled by the handle. Depending on the comfortable wrist angle, the user can switch between the two postures.
B. Shock absorption can be provided by the design of the handle.

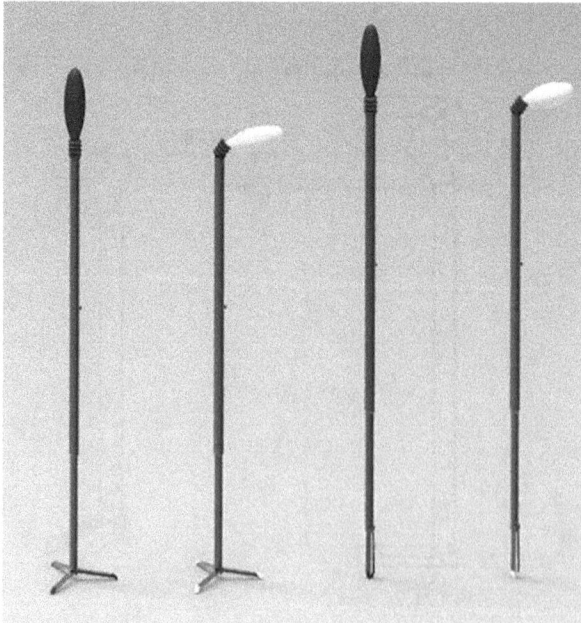

FIGURE 3.24 Walking stick concept 3.

C. As per the user's preference, the base can be switched from one leg to three legs depending on the surface.

D. An accessory-like look is given by the triangular form of the stick. This is in contrary with the circular form, which is associated with a walking stick used for medical aid.

E. Depending on the context of use, there are four combinations the user can adjust as per their comfort.

F. The body of the walking stick can be made from GFRP (Glass-Fibre Reinforced Polymer), also known as fibreglass. Although in absence of GFRP, Acrylic may also be used to make the stick lightweight and, thus, easily portable.

G. Mechanical advantages like high strength, water resistance and low weight-to-strength ratio can be provided by GFRP. This is also available with Acrylic Sheets.

H. The look of GFRP suits the triangular form of the stick and projects an over-all premium aesthetic to the walking stick. The same can be enforced using Acrylic Sheets as well.

The dimensions of the proposed design are taken as follows:

A. The handle length is 100 mm, which includes 90-mm palm length of 95th percentile male and an extra 10 mm for hand dynamicity.

B. The handle diameter is proposed to be 60 mm, which is slightly greater than 95th percentile male inner grip diameter enabling better power grip.

C. The Stick length is adjustable from 800 mm waist height of 5th percentile of the females to 1050 mm waist height of 95th percentile of the males.

The proposed dimensions may be altered looking at the accessibility of the material to fabricate the walking stick.

3.5 DISCUSSION AND CONCLUSION

The study follows the design thinking methodology for creation of ideas that solve the issues with the current designs of walking aid for the elderly, specifically in the Indian context. The comparison of ideas and current designs in the market on the basis of various qualities is also carried out. Figure 3.25 shows existing and proposed designs of walking sticks, which are then compared based on minimised Likert scale.

Table 3.4 compares the Proposed Designs with Existing Designs. The features of the designs are evaluated on 0–2 scale (minimised Likert scale). The definitions of the scores are explained below:

0 – No presence of the feature
1 – Weak presence of the feature
2 – Strong presence of the feature

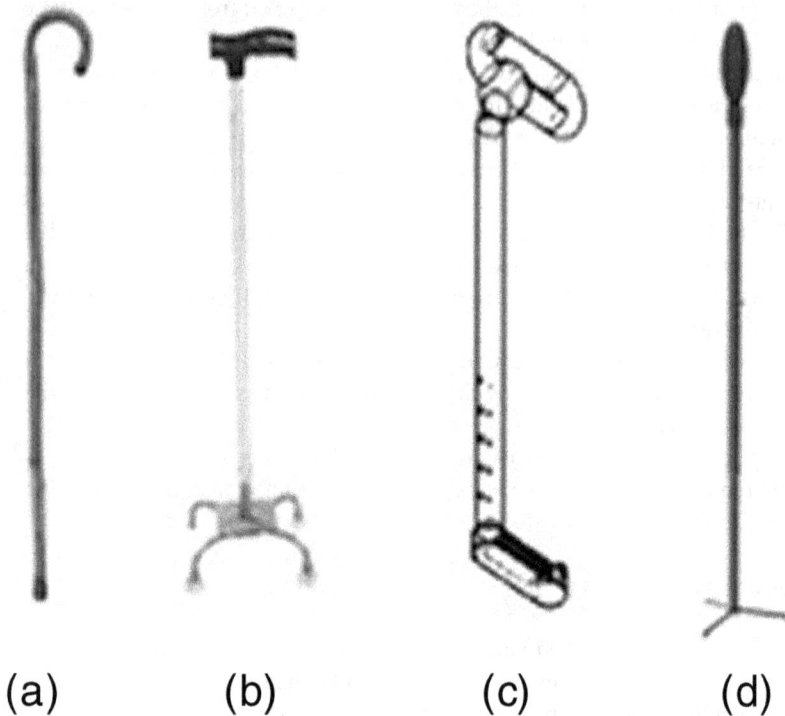

(a)   (b)   (c)   (d)

FIGURE 3.25 The various types of walking sticks for comparison. (a) Existing design 1 (E. D. 1) (b) Existing design 2 (E. D. 2) (c) Proposed design 1 (P. D. 1) (d) Proposed design 2 (P. D. 2).

TABLE 3.4

Design Comparison of Proposed with Existing Products

Designs	Balance	Grip	Ergonomic	Adjustability	Storage	Illumination	Aesthetic	Total
E.D 1	2	0	1	0	2	0	1	6
E.D 2	2	1	2	2	1	0	1	9
P.D 1	2	2	2	2	1	2	1	12
P.D 2	2	2	2	2	1	0	2	11

E.D. – Existing Design, P.D. – Proposed Design

It is found that the proposed designs score more over the existing designs based on the minimised Likert scale.

The study helped in creating walk aids that cater to three needs of the elderly. First, to reduce stress on palm (Design Concept 1), second, multipurpose usage (Design Concept 2) and third, aesthetics and accessory-like appearance (Design Concept 3). The target audience includes both the urban and rural elderly, ideally elderly with previous experience using a walking aid. The aim of the walking aid is to not only fulfil the needs of the elderly but also extends to the needs of people involved in their lives.

REFERENCES

1. Debkumar Chakrabarti, *Indian Anthropometric Dimensions for Ergonomic Design Practice*, National Institute of Design, 1997.
2. P. Mukhopadhyay, *Ergonomics for the Layman*, CRC Press, 2019.
3. Michael J. de Smith, *Statistical Analysis Handbook: A Comprehensive Handbook of Statistical Concepts, Techniques and Software Tools*, The Winchelsea Press, Drumlin Security Ltd., Edinburgh, 2018.
4. Hugh Young, *Statistical Treatment of Experimental Data*, McGraw Hill, 1962.
5. Siddhu Kulbir Singh, *Methodology of Research in Education*, Sterling Publisher, New Delhi, 1992.
6. S. P. Sukhia, P. V. Mehrotra, *Elements of Educational Research*, Allied Publisher Private Limited, New Delhi, 1983.
7. Martyn Denscombe, *The Good Research Guide*, Viva Books Private Limited, New Delhi, 1999.
8. James Garratt, *Design and Technology*, Cambridge Press, 1996.
9. P. John Williams, Kay Stables, *Critique in Design and Technology Education*, Springer, 2017.
10. Norman A. Donald, *Design of Everyday Things, Expanded and Revised*, Basic Books Publishing, 2013.
11. Clark Moustakas, *Heuristic Research: Design, Methodology, and Applications*, Sage Publications, 1990.
12. Waldemar Karawowski, *International Encyclopedia of Ergonomics and Human Factors*, CRC Press, Vol. 3, 2003.
13. Karl H. E. Kroemer, *Extra-Ordinary Ergonomics*, CRC Press, 2005.
14. S. Pande, A. Kenjale, A. Mathur, P. D. A. Kumar, B. Mukherjee, *Re-design of the Walking Stock for the Elderly Using Design Thinking in Indian Context*, Innovative Product Design and Intelligent Manufacturing Systems, Springer Singapore, pp. 29–39, 2020.
15. Chan King-Fai, Dongguan Shi (CN), "Multifunctional Walking Stick", May 4, 2006, Pub. No.: US 2006/0090783 A1.
16. Frank A. Lisi, *Combined Walking Stick and Bag*, Philadelphia, PA, August 6, 1940.
17. Alfred A. Smith, "Walking aid cane", 13114 Margate St.,Van Nuys, Calif. 91401, August 30, 1977.
18. W. Wu, L. Au, B. Jordan, T. Stathopoulos, M. Batalin, W. Kaiser, J. Chodosh, *The Smart Cane System: An Assistive Device for Geriatrics*, Institute for Computer Sciences, Social Informatics and Telecommunications Engineering (ICST), 2009.
19. Y. V. Wong, S. H. Yang, "Systematic Review on Handles of Canes for the Elderly's Daily Ambulation", *International Institute of Social and Economic Sciences*, 2017.
20. Robert Holliday, *Walking Stick Papers*, Al Haines, 2018.
21. Z. Tutuncu, A. Kavanaugh, "Rheumatic Disease in the Elderly: Rheumatoid Arthritis", *Rheumatic Disease Clinics of North America*, Vol. 33, No. 1, pp. 57–70, 2007.
22. M. Van der Esch, M. Heijmans, J. Dekker, "Factors Contributing to Possession and Use of Walking Aids among Persons with Rheumatoid Arthritis and Osteoarthritis", *Arthritis Care and Research: Official Journal of the American College of Rheumatology*, Vol. 49, No. 6, pp. 838–842, 2003.

4 Technology Intervention in the Designs of the Walking Stick

The previous chapter detailed the design thinking concepts in order to identify the bottlenecks in the re-design issues of the walking sticks in the Indian context. The anthropometry, was thus, specifically collected from the age group of the elderly (i.e., more than 60 years only) from both males and females. Based on the dissemination of these details, design thinking pedagogy helped achieve three design concepts. These design concepts are fabricated to build a prototype and then tested among users to collect feedback for the walking sticks so developed.

The fabrication issues deal with the choice of raw materials to build the prototype and the kind of technological intervention intended with each of the walking sticks. This also includes the nature of the colour scheme that needs to be adopted so as to make the walking stick more plausible for the elderly. This chapter provides a comprehensive study of the colour schemes adopted, fabrication process and the technological intervention in the walking sticks. Further, a user study to collect the feedback of the prototypes is also presented, and their correction factors, as suggested in the feedback interviews, are incorporated in the designs.

Three designs of the walking sticks are fabricated. Material selection, concept of design, electronic circuits implementation, physical limitations of the user and psychological interpretations of the user were kept in mind, while fabricating. Prerequisites for the fabrication process include the following:

- Material must be strong enough to support the weight of user. However, at the same time, it should be light enough so that the elderly can carry it conveniently from one place to another. This is to ensure that the walking stick does not add any kinematic load on the elderly user.
- Non-conducting material must be used, because electronic circuits will be placed inside stick. In order to house any electronic circuitry, wires will run around carrying a small voltage connected to a battery. Though so, even if the wires, due to any reason, are left open, the walking stick material should not be conducting to avoid any electric fatality for the elderly user.
- Mechanical stability to the fabricated stick while using it. This is of utmost importance because it serves the main purpose of using a walking stick.
- The walking stick should have the capability to house electronic circuit inside it. This should be ensured that there is no water seepage that may lead to any short circuit or cause any harm/fatality for the user. To augment the requirements, a proper and correct choice of colour scheme for the walking sticks also needs to be ensured. The colour scheme should be such that it is visible and identifiable from a certain distance that an elderly

DOI: 10.1201/9781003414957-4

is using a walking stick support. However, it should not even project the elderly as if he/she is in a state of medical exigency. This might affect the self-consciousness of the elderly, and they might refrain in using the walking stick ever.

4.1   COLOUR SCHEME

Selecting a colour scheme for the walking sticks required a study of colours and their aesthetic and psychological meanings interpreted by the Indian elderly [1–6]. Understanding the effects of colours on the physical eye and their emotional and subconscious impacts on the user gives an insight into the fact leading to choice of which colour would be ideal for a walking stick.

Colours are divided into Warm colours (comprising of Red, Orange and yellow) and Cool colours (comprising of Blue, Green and Purple). White, Black and Grey are categorised as Neutral colours.

- *Warm colours* are interpreted as vivid and energetic. They liberate the strong feelings of warmth and summer. They are usually associated with emotions of anger and alertness. Some specific details of each individual warm colour are mentioned below:
  1. Red colour is found to stimulate hunger and demands attention in a powerful way. Thus, the traffic signals use red light for stopping the vehicle at road crossings. Any emergency condition is also painted in red colour.
  2. Orange usually represents creativity, youth and enthusiasm. This colour simply attracts attention.
  3. Yellow is interpreted to represent happiness, hope and spontaneity.

  But as the product is made to be used by the Indian elderly, the use of solid warm colours is avoided as they might be interpreted as a religious sign. For example, saffron or orange colour is a close associate with Hinduism religious beliefs.

- *Cool colours* are supposed to reflect calm and soothing behaviour or situation. They liberate out the feelings of spring and winter.
  1. Blue colour is associated with calmness, intelligence, therapy and meditation. The logos of most institutions are shades of blue as it radiates feelings of trust, dependability and professionalism.
  2. Green colour represents nature, growth and stability. It brings a sense of visual balance and radiates feelings of relaxation. It also refers to a fertile and prosperous landmass.
  3. Purple colour represents royalty and luxury. It is also associated with spirituality and mysticism. It has the energy of warm colours and the calmness of cool colours.

  In the Indian context, cool colours are generally seen to be associated with a masculine nature.

- *Neutral colours* consist of black, white and grey.
  1. Black colour represents sophistication and power. It represents power and exclusiveness. When used in proper combination with other colours, it can radiate various emotions. This is one of the reasons that for showcasing a mass-scale protest against some decision, the protestors tend to put a black ribbon around the hand or the wrist.
  2. White colour is considered to be simple and minimalist. White is sometimes also associated with a certain level of purity.
  3. Grey (combination of Black and white colour) colour depicts formality, responsibility and maturity. It represents a strong character.

Different shades of colours are interpreted with different meanings. Lighter shades can get the attention immediately, whereas darker shades can seem more natural and mature. Using neon shades can give the colours a catchy effect, but it also seems to be quite artificial.

The ideal choice of colours for the walking stick is chosen from neutral colours as they are not associated with any specific gender. Black and muted silver are silent colours and, hence, they can easily merge into the lifestyle of the user. Black colour can also be seen from a distance and, thus, adds to illumination effect and visibility/attention.

4.2 TECHNOLOGICAL INTERVENTION THROUGH PROCESS FLOW

The design thinking process also has a very important tool for analysis known as the journey map [7–11]. A journey map is a visualization of the process that a person goes through in order to accomplish a goal. The journey map is an effective tool to compile the series of user actions into a timeline. Thereafter, the timeline is populated with various user thoughts and emotions in order to create a narrative. This narrative is condensed and polished, ultimately leading to a visualization of a design concept for implementation.

Most journey maps follow a certain format: at the top, a specific user or a specific scenario is detailed. The corresponding expectations or goals are placed in the middle area. The various high-level phases, which are comprised of user actions, thoughts and emotions, constitute the middle section. At the bottom, the different takeaways comprising opportunities, insights and internal ownership are explained in detail.

The key elements of a journey map are as follows:

1. Actor or the persona, on whom the complete journey map is based. The actor/actors are also the users of a product, users using a UI/UX, users confined in a space, users or actors in case of human–machine interaction (HMI) or human–computer Interaction (HCI), etc.

    The design is meant to satisfy the dire needs of the actors or the users. The target user group has to be precisely identified for forging ahead with a Journey map representation.

2. Scenarios and expectations describe the situation that the journey map addresses and is associated with an actor's goal or needs and their specific/categorical expectations. Scenarios can be real (for existing products and services) or anticipated – for products that are yet in the design stage or yet to be rolled out for commercial usage.

3. Journey phases provide organization for the rest of the information in the journey map (actions, thoughts and emotions). This is step-by-step movement from one task to other in order to accomplish a clearly laid out goal and objective. Each phase is jotted down with clarity so that one may identify the movement of the process they are engaged in.

4. Actions, mindsets and emotions pertaining to the actor are also critically important. The emotional details that the user bears throughout the journey are also mapped within each of the journey phases. Actions are the actual behaviours and steps taken by users. This component is not meant to be a granular step-by-step log of every discrete interaction. Rather, it is a narrative of the steps which the actor takes during that phase. Mindsets correspond to users' thoughts, questions, motivations and information needs at different stages in the journey. Ideally, these are customer verbatim from research. Emotions are plotted in a single line across the journey phases, literally signalling the emotional 'ups' and 'downs' of the experience.

5. Opportunities (along with additional contexts such as ownership and metrics) are insights gained from mapping. These are necessitated in order to provide the most optimised solution to the user.

The major advantage of a journey map includes a clear representation of the user expectations along with a mental model to fit all the requirements of the user. Also, the shared artefact resulting from the mapping can be used to communicate an understanding of the user or service to all the participants involved in the process. Journey maps are effective mechanisms for conveying information in such a way that it becomes memorable and concise and fosters a shared vision.

Based on the above, the fabrication process, technological intervention and user feedback are mapped into a journey map for clearly identifying all the pitfalls in prototyping. This is explained in the Process flow. There are some stages in order to accomplish the ultimate goal, which is technological intervention in the designed walking stick. Stages are presented in a table with separate columns. This is detailed in Figure 4.1.

The different stages of the process flow are described as follows:

I. Define
   • The first aim of 'Define' is to describe the different types of walking sticks present in the market. These include all the commercially used walking sticks of all price variations.
   • It is of utmost importance to identify which type of walking sticks are most preferred by the masses or general public (specifically, the elderly).
   • The next agenda is to identify the problems users face while using those walking sticks. Why is there a problem or why the problem persists

Scenario: Designing a Walking Stick for elderly which will have technological solutions in times of user requirement and emergency.
Expectations: Electronic components required to fulfill the technological solutions.

I. Define	II. Enquire	III. Check	IV. Test	V. Compare	VI. Select	VII. Installing the electronic components in design	VIII. Testing
1. Define present walking sticks 2. Define the problems with present sticks. 3. Define the Situations where the present sticks are lacking to provide the solutions. 4. Define the technological Intervened solutions to such problems.	5. Enquiring for required technological components which provide the desired output. 6. Selecting the components which provide the desired output.	7. Check whether the components fit in the space of designed sticks or not.	8. Test the electronic components whether they are fulfilling the requirement or not. 9. Testing the components individually and when all are connected.	10. Compare the components which provide the desired output and fit in the design, to sort out the best.	11. Select components which are best for design.	12. Fitting the components to walking stick according to requirement.	13. Testing whether the components are working or not.

FIGURE 4.1   The process flow of the technological intervention inspired from journey map technique.

needs to be introspected in depth. This is also included at this stage of the process flow diagram.

- The next pertinent question that arises is, what might be the technological solution to the walking sticks so that it solves the problem for the elderly?
- The proposed design solution should provide light to the users when it is dark and during emergencies such as power cut, using LED strips. This can be one of the proposed set of solutions.
- Emergency alarm and sending location to registered number were other solutions identified during times of emergencies, so that immediate assistance can be provided to the elderly in dire need.

II. Enquire
- To achieve desired technological solutions, enquiring and selecting the desired electronics and the associated circuitry is an important stage in this process.
- There are lots of electronic modules in the market which can give the same kind of output.
- In the case of microcontrollers, Arduino Series (Nano, Uno, Mega) and Raspberry Pi, etc. are all available for commercial and research use.
- In the case of GSM modules, in SIM Series 900A, 800L, 808, etc. are already available in the market.

- In the case of GPS modules Adafruit Fona SIM 808, NEO Series modules, etc. are already available in the market.
- In the case of power supply to the circuit, there are many different types of power banks or battery banks available in different sizes, weight and capacities. The choice for the context of walking stick would be less weight, longer life without charging and easy charging procedural capabilities for the comfort of the elderly.
- For the working of electronic modules, coding is required for interfacing the electronic components/sensors with the microcontroller selected.
- Enquiring about coding and programming language is also an important aspect.
- C++/C language coding is popularly used for deploying instructions to the microcontroller (e.g., for Arduino UNO, C/C++ are used).

III. Check
- Since the electronic components must are housed inside the walking stick, they have to be checked whether they can fit in the constrained space of the designed/fabricated sticks or not. A small space to incubate the electronic circuitry along with accommodating the power supply is a major challenge.
- In the second proposed design, Adafruit Fona SIM 808(GSM+GPS) module is proposed and used. This module has an integrated Global System for Mobile Communication (GSM) and Global Positioning System (GPS) module integrated into one chip. Due to size and space constraints, separate GSM module and GPS module units cannot be used.
- For the third proposed design solution, two models of integrating electronic circuitry are proposed. One with SIM 800L & NEO-6M (GPS) module and the other with ADAFRUIT FONA SIM 808(GSM+GPS) module. This in turn tests the efficacy of both using separate GSM and GPS modules and the integrated module for emergent situations while using the walking stick.
- In power supply, power bank of small size is always preferred.

IV. Test
- Before incorporating circuits inside the walking sticks, components must be tested individually whether they fulfil the basic requirements and needs of the users leading to the desired outputs.
- For testing purpose, software coding can be done for modules individually. This is in tandem with the Universal Design principle of Chunking or Top-Down approach.
- Due to bugs in coding, sometimes the modules are not able to generate the desired output.
- Sometimes the GPS and GSM modules did not receive signals properly from the satellites and towers due to the low power supply to the GSM and GPS modules. Sometimes due to external factors, like cloud cover,

cemented roofs which hinder penetration of the electromagnetic eaves, etc., the modules did not generate the desired output.
- While testing, it was found that the battery connected to SIM 800 L modules must be charged separately using TP 4056 module. The battery has to be removed from the circuit and has to be charged separately. This is a major constraint in using this module.
- TP 4056 module has B+, B− icons in it. The + (positive polarity) and − (negative polarity) of the battery must be connected respectively to charge the battery.

V. Compare
- To sort out the best, the various electronic components and circuits must be compared with each other. This gives an idea of the most-apt solution leading to low cognitive load on the elderly while using the walking stick.
- It is extremely important to solve the bugs in the software coding for interfacing the microcontroller. Using trial and error method, one can sort out the code used for the program.

VI. Select
- This refers to selecting various components which are best suited for the proposed design solutions.
- It also refers to selecting a particular set of codes best suited for the circuit application.

VII. Installing the electronics components in design
- After selecting the components, the components are installed into the respective walking sticks according to the requirements.
- Installing various components and connecting the components to ensure system-level integrity is an important stage. The various modules designed are now connected in a bottom-up approach so that the system can work cohesively.

VIII. Testing
- After installing all the components, testing is necessary whether the connections made are correct or not.
- Coding for the total circuit is an important step of the testing procedure.
- Testing also includes whether the stick is working properly or not after aggregating the electronic circuitry, which is an important stage in this process.

Based on the Process discussed above, the various electronic components are identified which can be retrofitted into the design to fabricate the prototypes of the walking stick with technological intervention. The different components used are as follows:

1. Arduino UNO – It is a microcontroller board [12]. It has 14 digital input and output pins, 6 analogue inputs, a 16 MHz quartz crystal, a USB connection, a power jack and a reset button. It can be connected to a computer or laptop

    for interfacing with other components and for deploying the software codes in it. It requires coding on a platform called Arduino IDE for functioning, which can be bunted into it through a computer or a laptop.

2. Adafruit FONA 808 Mini GSM + GPS – It is a cellular phone model with an integrated GPS that provides location tracking [13]. It can connect to a global GSM network with any 2G SIM. It can be operated using a microcontroller.

3. MPU 6050 – It is a sensor that contains an accelerometer and gyrometer in a single chip [14]. An accelerometer sensor is a tool that measures the acceleration of any body or object in its instantaneous rest frame. They can also be used to measure seismic activity, inclination, machine vibration, dynamic distance and speed with or without the influence of gravity.

    Gyro sensors or gyrometers, also known as angular rate sensors or angular velocity sensors, are devices that sense angular velocity. In simple terms, angular velocity is the change in rotational angle per unit of time. Angular velocity is generally expressed in deg/s (degrees per second).

    In the circuit implementation, the features of this sensor measure the roll, pitch and yaw. It can also be operated by using the Arduino UNO microcontroller.

4. SIM 800L – It is a GSM module [15]. It can connect to a global GSM network with any 2G/3G SIM. It can be operated using microcontroller.

5. NEO-6M – It is a GPS module [16]. It is used to track location in open space. It connects to satellite signals for providing details of location. It can be operated using a microcontroller.

6. Push button – A push button switch is a small, sealed mechanism that completes an electric circuit when you press it. When it is ON, a small metal spring inside it makes a contact with the two wires, allowing electricity to flow.

7. SPST switch – The Single Pole Single Through (SPST) is a basic ON/OFF switch that just connects or breaks the connection between two terminals. The power supply to a circuit is switched by the SPST switch. When the switch is open or off, then there is no current flow in the circuit.

8. Buzzer – A buzzer or beeper is an audio signalling device, which may be mechanical, electromechanical, or piezoelectric (piezo for short).

9. LED strip – A Light Emitting Diode (LED) strip is used to provide adequate luminance at night with minimal power supply from battery or the source connected.

Based on the above Process Flow, the detailed fabrication of each walking stick prototype and the technological intervention is carried out, which is explained in the subsequent sections.

The existing walking sticks are solutions for providing only physical support to the elderly while walking. The purpose of a walking stick can be enhanced with technological interventions.

The proposed features are:

- Lighting solution
- Location Tracker for cases of emergency. This may be implemented in both ways namely manual and automatic.

These specific features were selected because of their relevance. Only two features were selected because too many features would increase the cognitive load on the user and also its circuit implementation would require more space on the stick, which is a stricter limitation.

Technology Intervention or Location-Tracking circuit is initiated with two triggers. One is button-operated and the other is sensor-operated.

- *Button Operated:* This case is initiated by the elderly user himself or herself under the condition, when they are feeling uneasy or feel for a need of medical emergency. The button needs to be pressed for a long duration till a buzzer sound. The buzzer sound alerts the immediate people in their vicinity for providing any support. The Arduino UNO is turned on, and it activates the GPS module to search for network signals. If the signals are not found, the Arduino will reactivate the GPS module. If the signals are found, then the location co-ordinates will be obtained from the module and by using GSM module they are sent in the form of a Google Maps link to the registered phone numbers in the contact list of the elderly's mobile phone.
- *Sensor Operated:* This is used when there is a case of extreme emergency and the user is unable to press the button. If the elderly user feels extremely uneasy, there is a likelihood that they might fall. Falling can be taken as one of the emergent situations. The sensor MPU6050 measures the roll, pitch and yaw. If the values have crossed a certain set limit, the Arduino UNO is turned on and it activates the GPS module to search for network signals. If the signals are not found, the Arduino will reactivate the GPS module. If the signals are found, then the location co-ordinates will be obtained from the module and by using GSM module they are sent in the form of a Google Maps link to the registered phone number. Since there is a chance of mistake that the stick may fall down by accident and there was no actual emergency, the circuit gives a 20-second hold time. If push button is pressed within 20 seconds, emergency Short Message Service (SMS) will not be sent to the registered mobile number. Sometimes it also happens that the elderly might keep the walking stick at some angle for storage purpose. Thus, the threshold value is so set that it does not cross 60 degrees of inclination, which will activate the emergency system and the microcontroller.

The above implementation has the following limitations:

1. GPS module receives signals from satellite, whenever it is outside of a building.
2. So, the emergency SMS cannot be sent if the user falls down inside a building.

Similarly, GSM module must have enough signal strength in the area of user to send emergency messages through SMS or short message service on the cell phone.

4.3 WALKING STICK 1 CONCEPT: FABRICATION, FEATURES AND TECHNOLOGY INTERVENTION

Design Concept 1 of the Walking Stick is minimalistic and simplistic in approach. The design is proposed keeping in mind the fact that the average Indian elderly shall use it only as an aid for movements. Thus, the technological interventions are kept simple and the material used for fabrication is wood.

The key features of the prototype are as follows:

- Material used – Wood.
- Existing wooden walking stick is used as body.
- Three-pronged handle is carved out of a wooden block, at 120 degrees from each other and fixed in place of pre-existing single handle. Footrest is carved out of wood and inserted into body.
- For circuit implementation, grooves are made on the body, so that the circuit implementation does not disturb the initial design. These grooves act as passages for wires and for fixing the battery on the body. An inclined groove is made for placing the led strip, which is for lighting.
- Inclination is done to give direction to the light, and it does not cause visual disturbance to any person commuting in the opposite direction. Further, the inclination provided in the grooves for LED strips is directed towards the ground.
- Badminton racket grip cover is placed on the handle to provide a very good grip to elderly users.
- One pouch is nailed to the stick. The use of the pouch is to keep some small items preferably used by the elderly like mobile phones or some coins, etc. Wristband is tied to a stick with an aim that if the user wants additional gripping while holding the walking stick, it may be done using the wristband.
- Height of the prototype is kept approximately at 75cm, based on the design ideation provided for the same.

The fabricated prototype is clearly shown in Figure 4.2a–d, highlighting each of the specific part clearly.

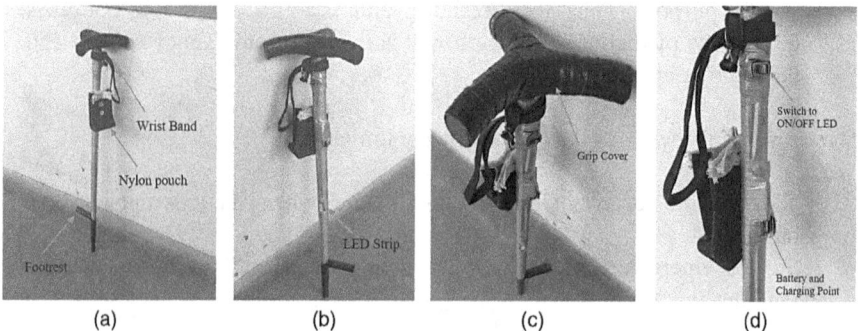

FIGURE 4.2 The prototype of the walking stick based on design concept 1. (a) Front view (b) LED strip mounted on walking stick (c) Handle with tri-grip of badminton racket (d) Charging point and switch.

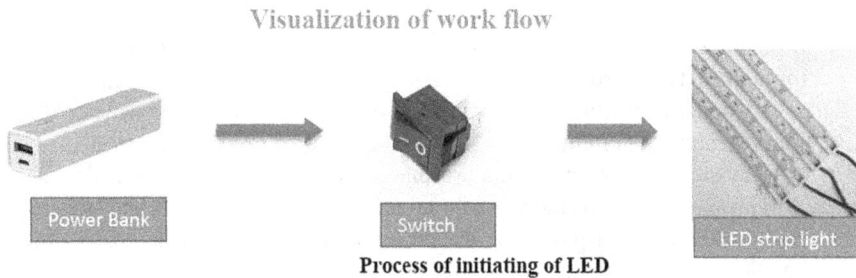

Process of initiating of LED

FIGURE 4.3   The workflow of the electronic circuit for operating the LED strip using a two-way switch.

The workflow of the circuit adopted for walking stick 1 is shown in Figure 4.3. Since this is a simplistic design, the electronics or technology intervention is kept simple. Through a single two-way switch, the power bank provides supply to LED strips for lighting. Also, the same power bank can be used for charging mobile phones, which can be placed conveniently in the pouch for recharging purposes. The battery or the power bank can also be easily recharged using a conventional mobile charging system or a simple Universal Serial Bus (USB) connector.

4.4   WALKING STICK 2 CONCEPT: FABRICATION, FEATURES AND TECHNOLOGY INTERVENTION

The walking stick design concept 2 is based on providing a firm grip to the user and also a footrest. Further, as the user study indicates that people keep the walking stick around themselves, it is important to provide a walking stick capable of standing alone without any support of wall or bedside, etc., as is needed for a conventional single-legged walking stick. However, while fabricating the same, weight is a major constraint, and the walking stick should not become too heavy for the elderly to use and should provide a firm grip. Since the cross-section of the walking stick is round, Polyvinyl Chloride (PVC) pipes are opted for construction as they are rugged, not fragile and lightweight. The key features of the fabricated prototype are stated below: -

- Materials used – PVC pipes (0.75-inch diameter), plastic balls, and vacuum-sealed boxes.
- A plastic ball section is used as a joint between the handle and the body.
- The frame or the body of the walking stick is made using PVC pipe. A slight curve around the neck of the stick is achieved by heating that end of the pipe and bending it to a required angle.
- Height adjustment rod with a pin mechanism is inserted inside through the base of walking stick. When the height needs to be adjusted, the pin needs to be pushed and adjusted to the hole for the required height.
- Holes are made along the main body or the frame of the walking stick for height adjustment.
- Filing of pipe for the flat shape for the height adjustment pin protruding is done so as to ensure that it is easily accessible.

- Base of the walking stick is made using lock and vacuum-sealed box. A hole is made into the lid of the box and the height adjustment rod is inserted through it. This portion is then sealed using glue that works strong enough even when the moisture in the ambience is very high. The box is waterproof for protecting the circuit and is strong enough to be used as a footrest.
- All the electronic components are placed inside the vacuum-sealed box. Wires are routed inside the pipes for connecting the button, LED and switch.
- Badminton racket grip cover is placed for smooth, soft and firm grip of the handle for the elderly users.
- Height of this prototype or the model ranges from 85cm to 100cm.

The fabricated prototype is clearly shown in Figure 4.4a–d, highlighting each of the specific part clearly.

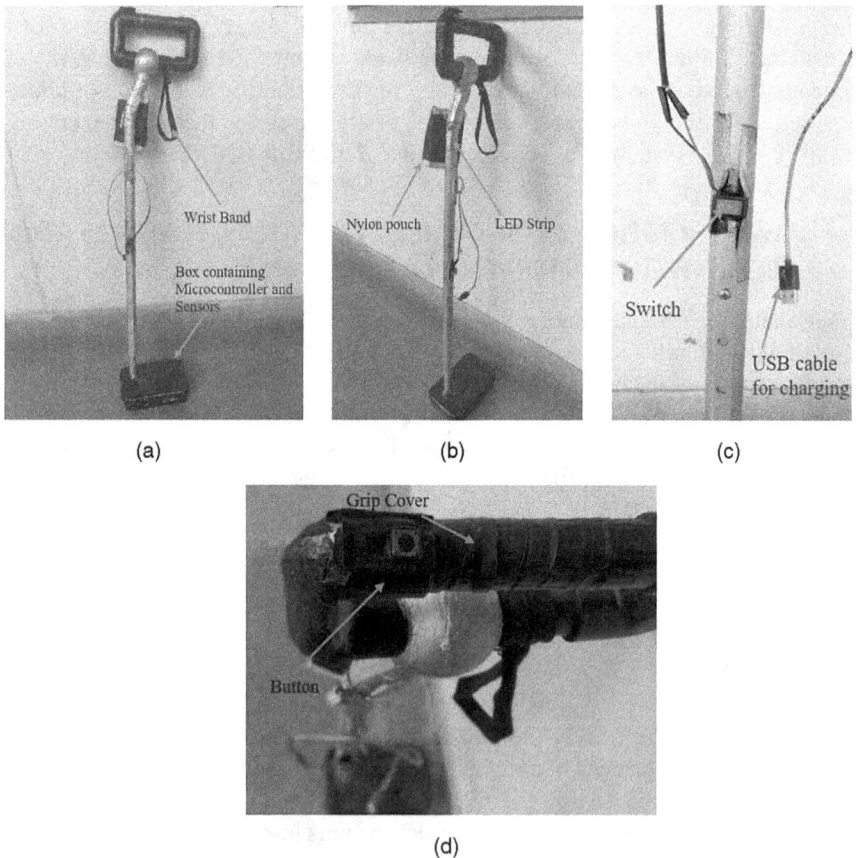

FIGURE 4.4 The various parts of the fabricated prototype of design concept 2 of walking stick. (a) Front view (b) LED light strip placement (c) Switch and USB cable for charging (d) Button on the handle for operating the electronic circuitry.

The second design was made using PVC pipes and a plastic ball for the junction of handle and body. The base was initially made from an inverted plastic basket, but due to technical limitations the base had to be an enclosed box for housing the circuit. A wooden box could be used for the base, but upon making the wooden box, it was found that it was very heavy and hence had to be changed due to the physical limitations of the user. Finally, a lock and sealed box is used as the base. It is light in weight and has sufficient space for the circuit to be kept and also provides waterproofing.

The technological intervention in the proposed prototype has three important functions. First, on a short press, the LED light strip is activated. Second, on a long press of button, a buzzer is activated and an emergency message with GPS location is sent to the close kin of the elderly. This is important to use when the elderly falls in some emergency situation and needs help. People in immediate vicinity get to know of it from the buzzer or alarm raised by the walking stick. Also, the relatives of the elderly shall get a message using SMS and the GPS location. The third function is activated when the elderly falls while using the walking stick. The Arduino sets a timer to check if the fall is by a mistake and in this time frame, if the stick is not placed back to its original position, an alarm is raised and an SMS with GPS location of the elderly user is sent to the relatives.

The above workflow is clearly shown in Figure 4.5a–c.

The detailed flowchart of the working mechanism of the electronic circuitry deployed in the Walking Stick as discussed above is illustrated in Figures 4.6 and 4.7. The two flowcharts have been deployed using C++ programming which is burnt in the microcontroller. The microcontroller then, based on the commands of the programs deployed, executes various functions as discussed above. In 2nd design, Adafruit Fona SIM 808 module is used for GSM and GPS modules. I2C protocol code must be used in parallel with main code in both the designs.

4.5 WALKING STICK 3 CONCEPT: FABRICATION, FEATURES AND TECHNOLOGY INTERVENTION

The Walking Stick Design Concept 3 is based on aesthetic use. Many users and elderly people avoid using a walking stick as it generates a temperament of medical sickness, despite the need for support in regular walking. For these individuals, the walking stick concept 3 is designed so as to make the stick look more like an accessory rather than a medical need and requirement. Keeping the same in purview, the fabricated walking stick has the following features:

- Materials used – Acrylic sheets, PVC pipes.
- Three stripes are cut and stuck in a triangular shape. At joints, Araldite is used for sticking. The same is done for height adjustment triangular shape.
- Holes are made to the outer part of the body or the frame of the walking stick using drilling. Height adjustment pin is inserted using spring and neutral pin of three-pin plugs.

Process of initiating of LED

(a)

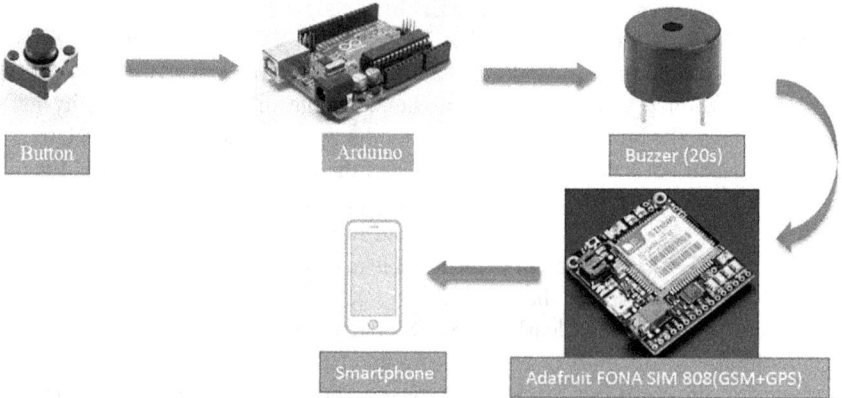

Process of initiating of buzzer and emergency sms manually

(b)

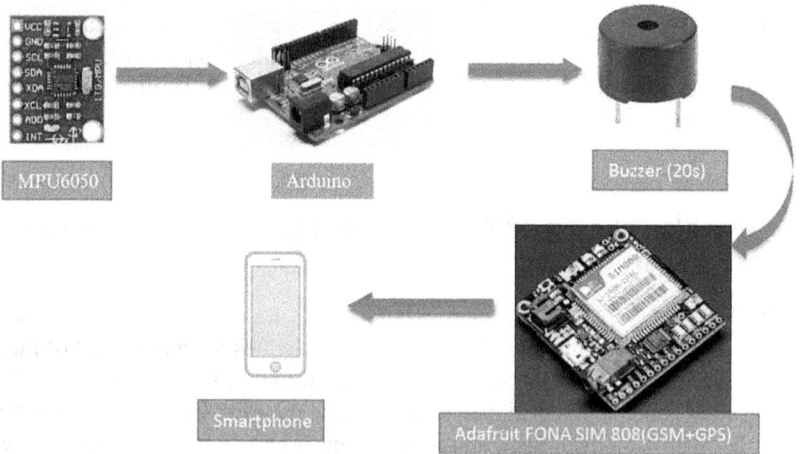

Process of initiating of buzzer and emergency sms

(c)

FIGURE 4.5 The workflow of the electronic circuitry for prototype of the walking stick design 2. (a) Function of a short press of button (b) Function of a long press of the button (c) Function in case the elderly falls with the stick.

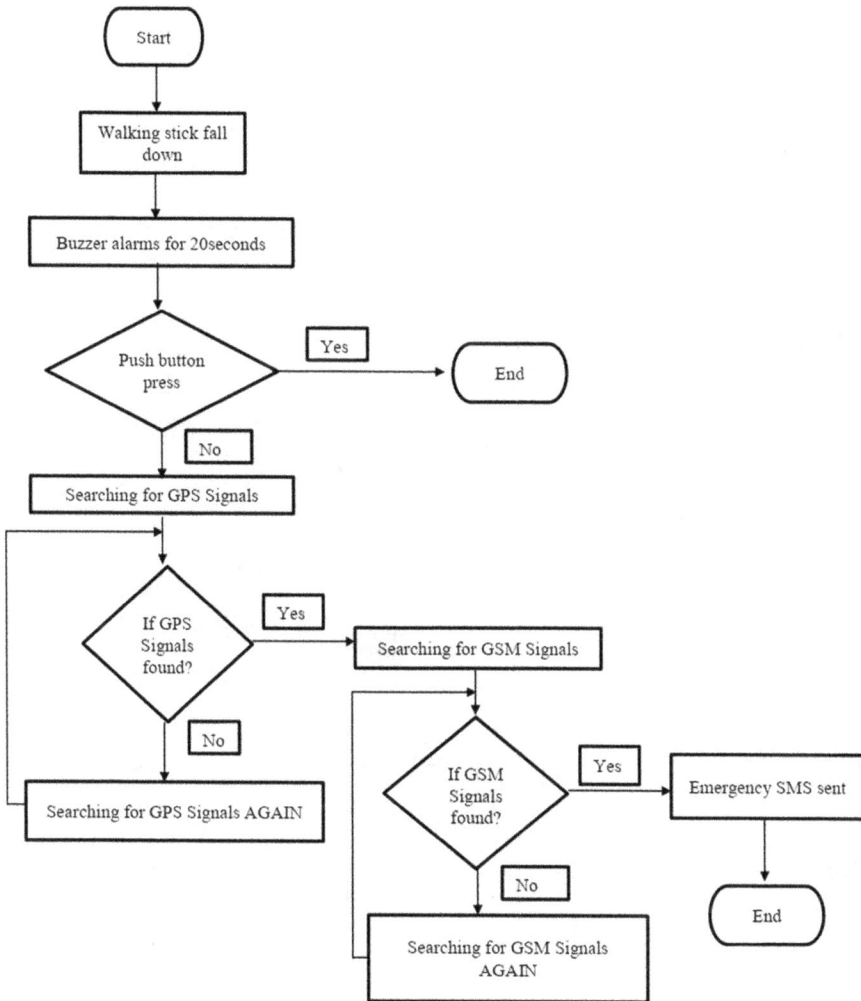

FIGURE 4.6 The flowchart of the commands to be executed for fall detection of the elderly.

- The handle of the walking stick is made using PVC pipe of 1-inch diameter pipe. It is interesting to note that contrary to the handle as prescribed in the design ideation, the handle is modified to being a U-shaped handle with a strong PVC pipe. If the handle is capable of changing position or can tilt as proposed in the design ideation, the elderly users do not get the confidence of a strong grip using the walking stick. Thus, it is replaced by a better and rigid gripping solution. Following the same mandate, a badminton grip and a wristband are further appended to improve the gipping for elderly users while holding the walking stick.
- Pipe is stuck to the three surfaces of the outer triangular body or the frame of the walking stick, as is closely evident in Figure 4.7c.

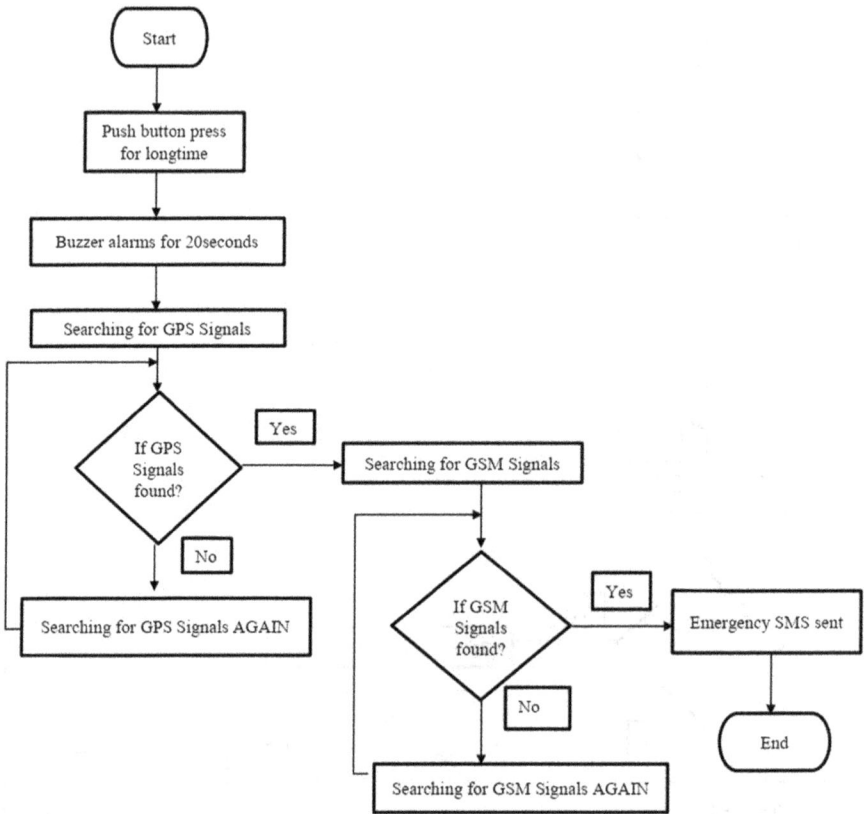

FIGURE 4.7 Flowchart of the commands to be executed in case an alarm needs to be raised by the long press of the button near the walking stick handle.

- The base of the walking stick is fabricated using small stripes of acrylic sheets and metallic hinge joint, and it is connected to the inner triangular frame or the body using a strong adhesive.
- Some holes are made for the electronic circuit wires to provide various connections.
- Arduino, power bank and battery are connected to the outer surface of triangular section, due to space constraints.
- Remaining components are placed inside the walking stick.
- Height of this model or the prototype ranges from 80cm to 105cm.
- Acrylic sheet is extremely light and when stacked and assembled is strong enough to resist the weight during a firm grip walking. Also, since individual acrylic sheets can be easily sheered or moulded in different shapes, keeping in mind the triangular cross-section, the material is also non-conductive to electrical circuits, making it a good insulator housing the circuitry for technological intervention. The detailed photographs citing the insights of the fabricated prototype are shown in Figure 4.8a–e.

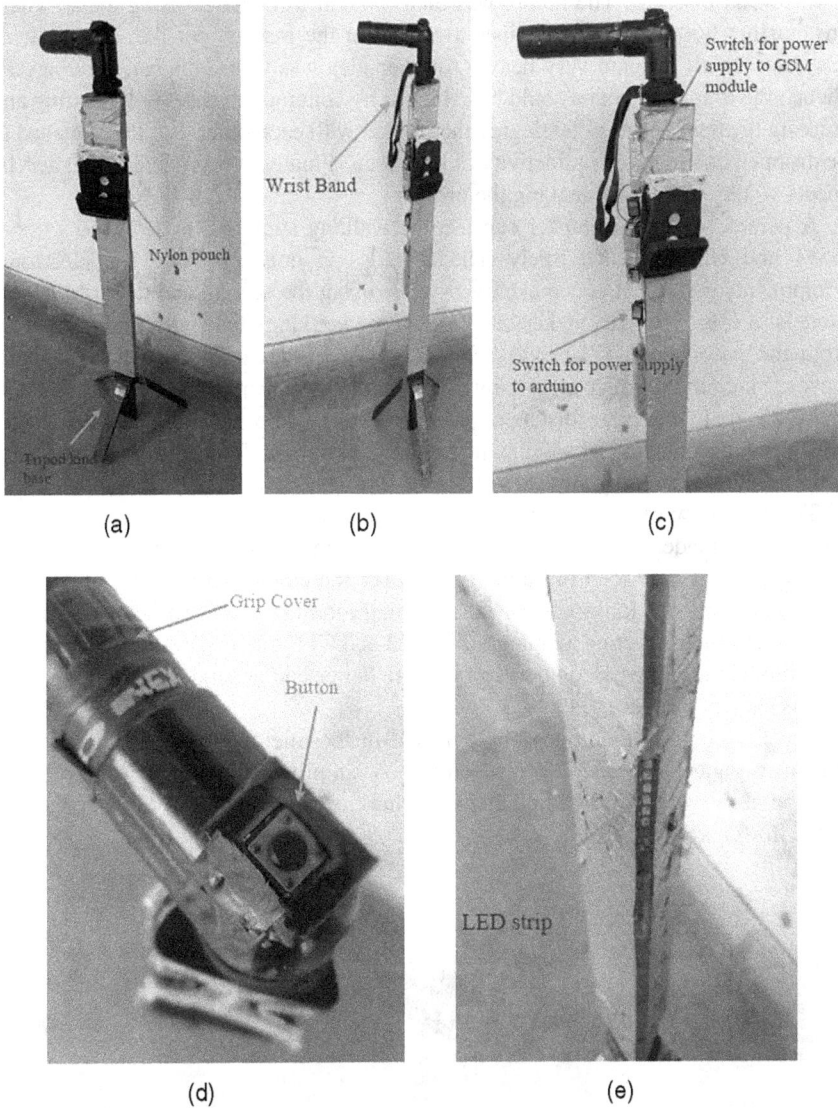

FIGURE 4.8 The various parts of the fabricated prototype of design concept 3 of walking stick. (a) Front view (b) Isometric view with the wristband (c) Switch location to activate GSM module and power supply (d) Button in handle to operate the circuitry (e) LED strip along the length of the walking stick.

The third design remains the most challenging of all the three proposed design solutions because of its triangular cross-section. It was suggested to be made using Glass Fibre Reinforced Polymer (GFRP), but working with this material was expensive. Wood was also not a feasible option as hollow triangular structure was difficult, and there was a risk of the model or prototype becoming heavy due to the wood material. The model was then thought to be made using metal. There are existing hollow triangular pipes available in the market, but they are made of cast iron which is again very heavy. Another idea was to use galvanised iron (GI) sheets; the triangular shape could be achieved by bending the sheet or by cutting and sticking the three strips at 60-degree inclination with each other. But this idea had to be dropped due to mild conductivity of the sheets. Finally, it was decided that acrylic sheets would be used for making the product.

A person with a pacemaker can use the walking stick that is designed without GSM and GPS modules freely. The first design does not have the electronic components that affect pacemaker. But while using the second and the third design models, a user with pacemaker has to keep the walking sticks at least 10cm away from the pacemaker [17] because the pacemaker is affected the electromagnetic waves. Since the product is walking stick, so the electronic components like GSM and GPS module are at a distance (at knee of user while walking) which does not affect the pacemaker. Thus, the proposed design solutions and prototypes are convenient to be used by any elderly.

The workflow of the technological intervention is almost similar to that of the intervention made for the second design concept implemented. All three functions are the same as described in Figure 4.5a–c. For the case when walking stick or the elderly carrying the walking stick falls, a small modification in components used is observed. Instead of using an integrated GSM and GPS module, a separate GSM and GPS module are integrated so as to facilitate better and accurate location tracking. This is shown in Figure 4.9.

The flowchart of the codes implemented in the microcontroller is the same as shown in Figures 4.6 and 4.7, respectively. A simplified function of long press to raise an alarm may also be implemented using the microcontroller as demonstrated in the flowchart of Figure 4.10.

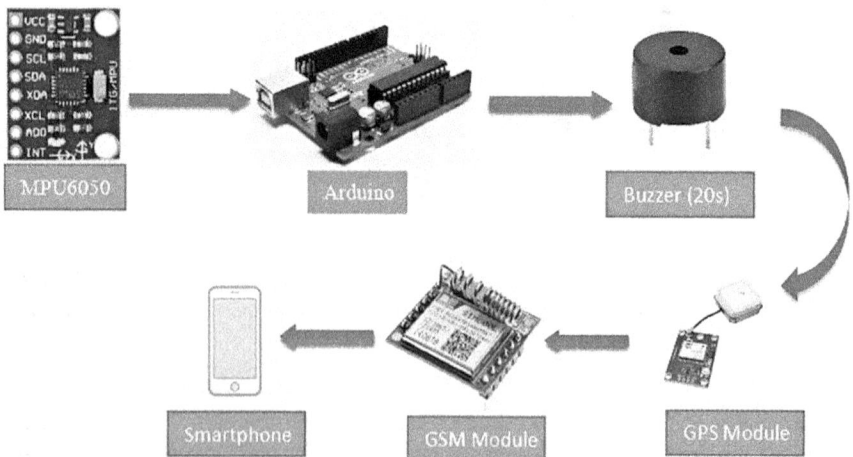

FIGURE 4.9 Process of initiating of buzzer and emergency SMS with GPS tracking when the elderly falls with the stick.

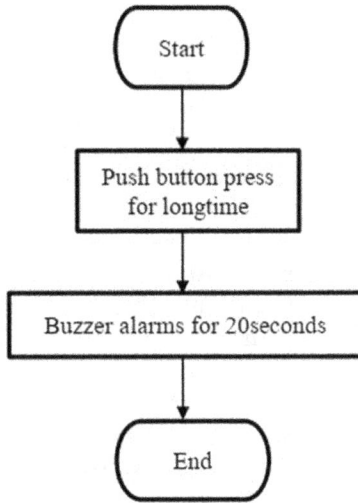

FIGURE 4.10   Simplified flowchart of raising an alarm from buzzer on a long press of the button near the handle of the walking stick.

C, C++ programming is used in Arduino IDE for microcontroller. Programming is done differently for the second and the third designs, because of the modules used. In the second design, Adafruit Fona SIM 808 module is used for GSM and GPS modules. In the third design, SIM 800 L module is used for GSM and NEO-6M module is used for GPS. I2C protocol code must be used in parallel with main code in both the proposed design solutions.

The prototypes can be developed in-house with minimum fabrication facilities. The need comprising of high-end machines is not substantiated for the fabrication of the proposed walking sticks. These simplistic solutions make them a suitable candidate for commercialization as well.

4.6   FEEDBACK OF THE WALKING STICKS THROUGH DESIGN THINKING APPROACH

Any design process is incomplete without taking the required feedback. Feedback helps improve the product cycle and its lifetime. The feedback also helps assess how the product is better than existing ones in the market. Feedback loop consists of both positive and negative feedback. In the context of the feedback of the products, positive feedback means that the technological solutions are acceptable to the users, and negative feedback would mean that the product needs further improvement and refinement.

It is interesting to quote an important observation of the feedback here. In the context of the Design Ideation of concept 3, the handle is proposed to be flexible in nature. This is in reference to Figure 3.24. The handle proposed in the ideation states that two gripping postures are enabled by the handle. Depending on the comfortable wrist angle, the user can switch between the two postures. However, when the feedback for the ideated concept was taken by the elderly, the feedback was negative. The feedback stated that if the handle itself is flexible, how the elderly will feel comfortable as the first body weight is laid on the handle itself. So it needs to be rigid and rugged so that the elderly can rely on it totally and place their body weight on it for mobility purpose. This feedback led to change in the ideated concept and the fabricated prototype of Figure 4.7 was corrected and tested for user study requirements and analysis.

The feedback for the prototypes developed is also taken by interviewing through a questionnaire. The participants were asked various questions, both qualitative and quantitative, to understand the impact of the prototypes created. The qualitative questions are highlighted in bold and italics. The quantitative questions are highlighted in bold and underlined as well. There are some notes added shown as bold plus italic plus underlined. This helps in understanding the user better. The questionnaire format is a detailed structured questionnaire which is executed through the process of interview as a part of Direct Observation and Analysis process.

There are two main types of users in the feedback scenario. *Primary users* are the elderly who shall be using the walking stick, and *Secondary users* are the ones who shall receive the SMS and GPS message. Thus, questionnaire covers both types of users and their feedback is also taken accordingly.

The details of the various questions in the questionnaire are as follows:

Primary User (Elderly User of the walking stick)

Age: …………………….. **(It is a Quantitative question which is asked towards the end of the interview session to avoid the Hawthorne effect. If this question is posed at the beginning of the interview itself, the likelihood that the responses received by the elderly shall be biased is likely to be very high.)**

Gender:……………………………. **(It is a Quantitative question which is marked or used for the data analytics and ideation of correction in the prototypes after the interview session of all the interviewees is completed. Sometimes the preferences of the male and female choices differ based on Gender, and so it constitutes an important point in the questionnaire.)**

1. What is your opinion on the handle of walking stick? *(It is a Qualitative Question where each response is noted. The feedback is taken for all the designs fabricated, from the elderly users.)*

2. On a scale how much comfort do you feel with the handle of the Walking stick? **(The question is to map the comfort level, which is a qualitative scale, with a Quantitative question using the Likert scale. This is asked for all the fabricated prototypes. The comfort level of handle refers to the improvement proposed over the existing conventional handles, which are discussed before. A scale of 1–5 is not chosen since it is not clear that some people might have injuries and so it might lead to inherent pain which cannot be mapped in a scale of 1 5. A scale of 1–10 is not chosen since this scale is chosen for more complex scales and since this is a focused group on elderly, a scale of 1–10 is avoided. Thus, a scale of 1 to 7 is included.)**

(Least) 1-----|-----|-----|-----|-----|----- 7 (most)

3. What is your opinion on the grip of the walking stick? *(This is a Qualitative question which takes into consideration how the grip is improved by improvement in the design of the handles loaded with the badminton grip and the wristband, taken together. Since the wristband and the badminton grip are placed for all the designs, the feedback is taken for all three fabricated prototypes.)*

4. On a scale how much comfort do you feel with the grip of the walking stick? *(This is a Qualitative question. The aim behind this question is to retrieve a quantitative data out of the qualitative feel or emotion of relief the elderly gets on using the improved gripping conditions. This is mapped in the Likert scale for better understanding.)*

    (Least) 1-----|-----|-----|-----|-----|----- 7 (most)

5. What is your opinion on the footrest part of the walking Stick? *(This is a Qualitative question. This is intended specifically for the footrest of the first fabricated prototype for design concept 1. In this prototype, a small footrest is extended near the base of the walking stick.)*

6. On a scale how much comfort do you feel with footrest design Walking stick? (This is a Quantitative question to map the comfort level of the use of the footrest as proposed in the fabricated prototype of the design concept 1. The data received can be plotted as a bar graph or a pie chart to find out how many elderly users find it comfortable enough to use the innovative footrest concept, as proposed.)

    (Least) 1-----|-----|-----|-----|-----|----- 7 (most)

7. What is your opinion on the lighting solution of the walking Stick? *(This is a Qualitative question. The lighting solution is inherently applied to all the fabricated prototypes. The solution aims to improve the visibility of the roads during dark or power-cut moments. Through this question, the elderly user places how they feel with such a solution in their walking stick product.)*

8. On a scale how useful is the lighting solution of walking stick? (This is a Quantitative question to plot the feelings or emotions of working with a lighting solution for the walking stick. The reason for the choice of the Likert scale from 1 to 7 is the same as for previous cases.)

    (Least) 1-----|-----|-----|-----|-----|----- 7 (most)

9. What is your opinion on buttons/switches provided on the walking stick? *(This is a Qualitative question. Interestingly, the button or the switch for the products is different. While a simple two-way switch is used for the simplistic and minimalistic fabricated prototypes of design concept 1, the other two designs have a press button which can be operated with a long press or a short press. The long press and the short press perform different operations as is evident from the flowcharts of the Figures 4.6 and 4.9, respectively. The short press of the button lights up the LED circuit for providing the lighting solution. Thus, the feedback is taken separately for all the three fabricated prototypes.)*

10. On a scale how much comfort do you feel with buttons/switches provided? (This is a Quantitative question to map the comfort level of using the buttons or switches provided on the walking stick, on the Likert scale rating. The reason for the choice of the Likert scale from 1 to 7 is the same as for previous cases.)

    (Least) 1-----|-----|-----|-----|-----|----- 7 (most)

11. What is your opinion on solution provided for emergency alarm? *(This is a Qualitative question. The opinion is sought on the alarm solution on the press of a long button as provided in the second and the third fabricated prototypes of the design concepts 2 and 3, respectively.)*

12. On a scale how useful is the alarm provided? <u>(This is a Quantitative question. The aim is to map the Likert scale rating of the qualitative inputs sought of the above question so as to map it for statistical treatment. The reason for the choice of the Likert scale from 1 to 7 is the same as for previous cases. This is also restricted to the second and the third fabricated prototypes of the design concepts 2 and 3, respectively.)</u>

     (Least) 1-----|-----|-----|-----|-----|----- 7 (most)

13. What is your opinion on emergency location tracking? *(This is a Qualitative question. This also meant for the second and the third fabricated prototypes of design concepts 2 and 3, respectively. The emergency location tracking is provided in the form of a google map link to the phone contacts as saved in the Subscriber Identity Module (SIM) card placed in the walking stick.)*

14. On a scale how useful is emergency location-tracking solution? <u>(This is a Quantitative question. The aim is to map the Likert scale rating of the qualitative inputs sought of the above question so as to map it for statistical treatment. The reason for the choice of the Likert scale from 1 to 7 is the same as for previous cases. This is also restricted to the second and the third fabricated prototypes of the design concepts 2 and 3, respectively.)</u>

     (Least) 1-----|-----|-----|-----|-----|----- 7 (most)

15. What do you feel about the walking stick? *(This is a generic Qualitative question. The opinion sought in this question is about all three walking sticks fabricated and their features improved through the various technological interventions provided. The suggestion from this query can be interpreted as a comparison of the three prototypes as well as valuable inputs to improve any section in the walking sticks, so proposed.)*

     Secondary User (The person who takes care of walking stick user or who specifically receives the google map links of the GPS location through the SIM inserted in the walking stick. On the secondary end, it can be a single user or multiple users, depending on how many user contacts should be sent the GPS location and SMS details which can be programmed in the microcontroller programming itself. Usually not more than five contacts are addressed or a priority contact list is provided to ensure that such emergency details are sent to them. Further, the role of the secondary users comes into picture for the second and the third fabricated prototypes of design concepts 2 and 3, respectively.)

16. What is your opinion on the emergency alarm and location tracking provided in walking stick? *(This is a Qualitative question. The inputs sought from the secondary users help improve the concept of location tracking of the elderly amidst an emergency-based situation. The secondary users can advise on how to improve the system for better comprehensibility or reduce complexity for the elderly user.)*

17. On a scale how useful is an emergency alarm and location-tracking solution? <u>(This is a Quantitative question. The aim is to map the Likert scale rating of the qualitative inputs sought of the above question so as to map it for statistical treatment. The reason for the choice of the Likert scale</u>

> **from 1 to 7 is the same as for previous cases. This is also restricted to the**
> **second and the third fabricated prototypes of the design concepts 2 and**
> **3, respectively.)**
> (Least) 1-----|-----|-----|-----|-----|----- 7 (most)

4.7 CONCLUSION

While taking the feedback, users are categorised into two types.

1. *Primary user* – Main user or the elderly using the walking stick.
   From the primary user, the feedback is collected twice.
   I. For assessing the design of the walking sticks. Design feedback is taken from 22 users.
   II. For assessing the technology intervention of walking sticks. Technology feedback is taken from 20 users.
2. *Secondary user* – The person who is the caretaker of primary user which is 21 in number. (Secondary user actually receives the emergency SMS with GPS location on their mobile phones respectively of the primary elderly user.)

Feedback given by the users is plotted on a pie chart. Likert scale is used to note down the response of the users and for projection of the quantitative data gathered from the users. Their response is also noted in terms of some words and numbers. These are as mentioned below:

- Very Good – 7
- Good – 6
- Above Average – 5
- Average – 4
- Below Average – 3
- Not Bad – 2
- Bad –1
- No Response – 0

It was interesting to note that after 15 users, the responses started reaching a saturation level. The feedback is analysed and explained using the pie charts. Here the pie charts are presented one by one for three designs of walking sticks. Among the people from whom the feedback is taken, some people state good, some state average, etc. So the responses are mentioned in percentages. Figures 4.11 to 4.16 show the feedback of the responses for the prototype of design concept 1 of the walking stick. Figures 4.17 to 4.22 show the feedback of the responses for the prototype of design concept 2 of the walking stick. Figures 4.23 to 4.28 show the feedback of the responses for the prototype of design concept 3 of the walking stick. Figure 4.29 shows the feedback of the secondary users on the technology intervention of the SMS received in design 2 and design 3. The qualitative responses for the feedback conducted on various design and fabricated prototype aspects were found to be in tandem with the idea behind the three ideated design concepts. Hence, they are not discussed here, as it is included in the ideation stage itself.

Among the 22 users who participated in the feedback interview, 77% users gave positive response in support of the modified design for the tri-junction handle. Among the ones who raised their apprehension on the feasibility of the handle design had issues like hand with only four fingers as one finger was lost due to an accident, small grip diameter as found in the anthropometry measurements, etc. This is shown in Figure 4.11.

Out of the participants who participated in the feedback interview, 50% users rendered positive response for the base design. The other users found that the base was slippery enough. Since the base was a conventional single-legged base, it was

1ST DESIGN HANDLE

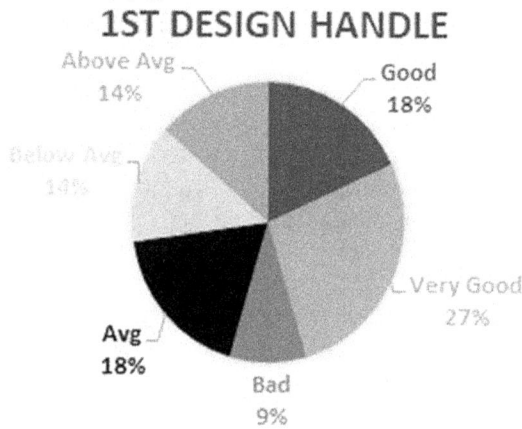

FIGURE 4.11   The feedback of the handle of the 1st design.

1ST DESIGN BASE

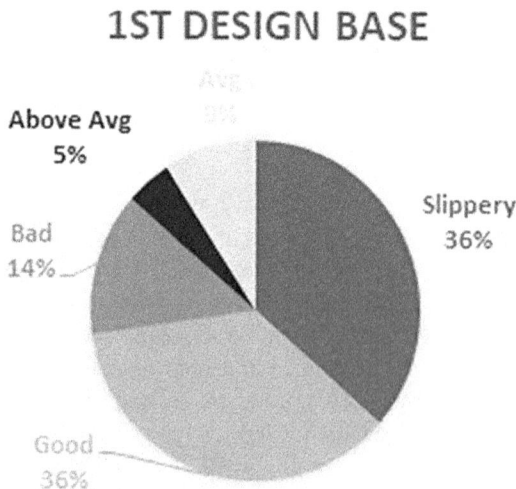

FIGURE 4.12   The feedback of the base or footrest of the 1st design.

found that people with walking stick prefer using a stronger base which can allow the walking stick to stand alone. This is shown in Figure 4.12.

Out of all the participants in the feedback interview, 64% of the elderly users appreciated with a positive response for footrest part walking stick. The additional footrest placed near the lower part of the walking stick was found to be comfortable while walking and provided the rest to one of the legs with due support of balance to the elderly user. The pie chart of the responses is shown in Figure 4.13.

Interestingly, 100% of the elderly users who participated in the feedback interview approved the lighting solution provided for the design concept 1 fabricated prototype. They gave a positive response to the lighting solution, which shows that the proposed solution addresses a dire problem for elderly users, who use a walking stick. The pie chart diagram plot of the responses made by the participants is shown in Figure 4.14. Not a single participant raised an apprehension towards the design solution provided.

1ST DESIGN FOOTREST

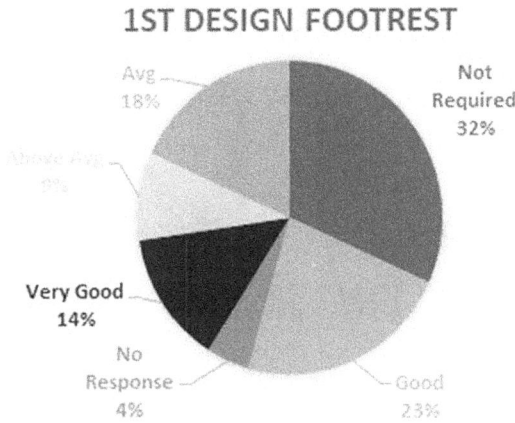

FIGURE 4.13 Feedback of 1st design footrest.

1ST DESIGN LIGHTING SOLUTION

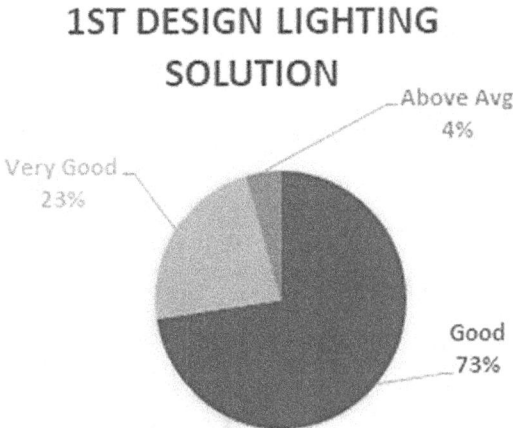

FIGURE 4.14 Feedback of 1st design lighting solution.

From the responses collected in the feedback, 45% of users appreciated the presence of switch provided in walking stick for lighting solution. Those who found it uncomfortable, which constitutes 35% of the elderly users participants have placed their suggestion that the switch is needed, however, it should be placed somewhere close to the handle. This was incorporated in the fabricated prototype as shown in Figure 4.2. After the correction in the design, the rest of the 35% of the participants also approved the concept. The pie diagram of the response of the presence of the switch is shown in Figure 4.15.

Encouragingly, 76% of the participants or the elderly users gave positive response to 1st design of the walking stick. They found it to be ergonomic and easy to use with better technological solutions to ease their overall quality of life. This variation of the data is plotted in the pie diagram, as shown in Figure 4.16.

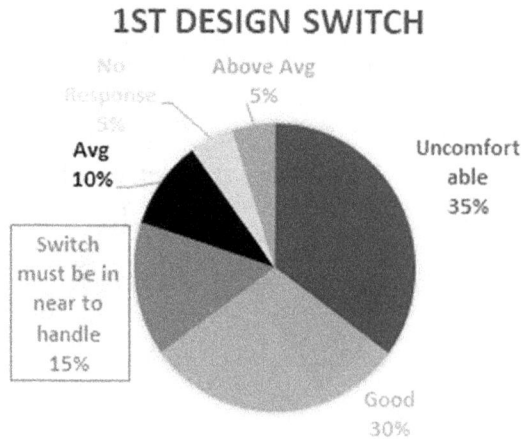

1ST DESIGN SWITCH

No Response 5%
Above Avg 5%
Avg 10%
Switch must be in near to handle 15%
Uncomfort able 35%
Good 30%

FIGURE 4.15 Feedback of 1st design switch.

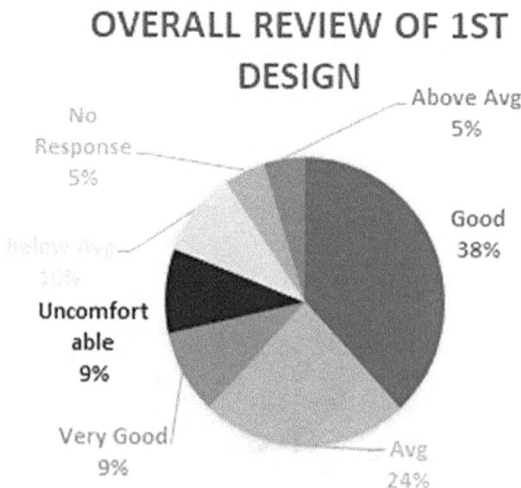

OVERALL REVIEW OF 1ST DESIGN

No Response 5%
Above Avg 5%
Good 38%
Below Avg 10%
Uncomfort able 9%
Very Good 9%
Avg 24%

FIGURE 4.16 Feedback of 1st design overall review.

The rectangular curve handle for design concept 2 was much appreciated by the participants in the survey. Out of all the participants, 95% of the elderly users favoured the improved handle of the second design. They found it to an extent a better support during walking. This is shown in Figure 4.17.

Seventy-seven per cent of the participants or users of the feedback survey placed their consent to the fact that the base grip of the walking stick was improved with the design solution provided. This base offered a very comfortable footrest to stand comfortably during a walk. Further, since the base can be left standalone without any support from the bedside or on the ground, it provides better stability to the users also. The feedback responses as plotted on the pie chart are shown in Figure 4.18.

2ND DESIGN HANDLE

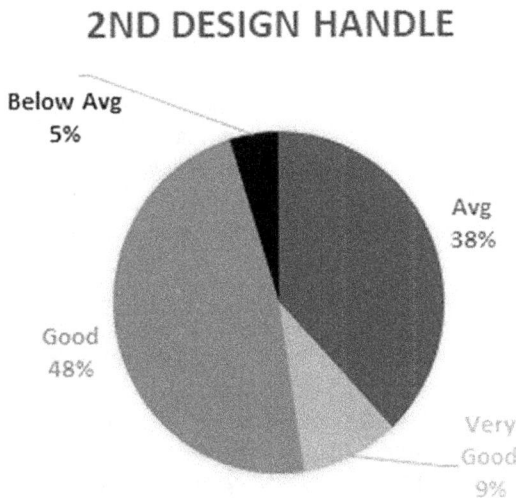

FIGURE 4.17 Feedback of 2nd design handle.

2ND DESIGN BASE

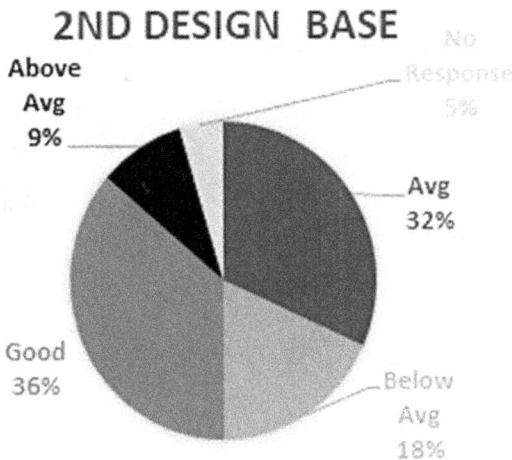

FIGURE 4.18 Feedback of 2nd design base.

Five per cent constituted of those elderly users or participants who have recorded a no response as feedback or were not able to judge as they were not able to compare their walking stick with the base of the proposed second design.

Ninety per cent of the elderly users were of the opinion that the lighting solution in the second design is appropriate and much needed. The lighting solution in the fabricated prototype addresses the dire problem of walking with a stick at night or during power failures at night. This feature adds independence to the elderly users to walk without constraints even in the dark or night-time. The pie chart figure is shown in Figure 4.19, marking the responses made by the participants of the feedback survey.

As many as 76% of the total participants in the feedback user study were of the opinion that the button to turn on the electric circuit in the walking stick is comfortable to use. They gave a positive response to the button provided in the walking stick. This is plotted in Figure 4.20, which also shows that 5% of the participants could not respond appropriately.

Eighty-five per cent of the users or participants favoured the presence of a switch provided in walking stick, which is of push button type. The switch was found to be ergonomic in tandem with the fabricated prototype. The responses are plotted in Figure 4.21.

Eighty per cent of the users gave positive response to the emergency alarm and SMS solution of walking stick as proposed in the fabricated prototype of the design concept 2. Thirty per cent of the users find it to be very encouraging. The 30% users who voted in favour of the solution provided have kept themselves updated with the technological advances taking place. They are also users of smartphones and, thus, can understand the necessity of sending an SMS along with GPS location during emergent situations. The various recorded responses are plotted in pie chart and shown in Figure 4.22. The 5% of the users who voted it to be below average are those who either do not have mobile phones or are still using conventional keypad phones.

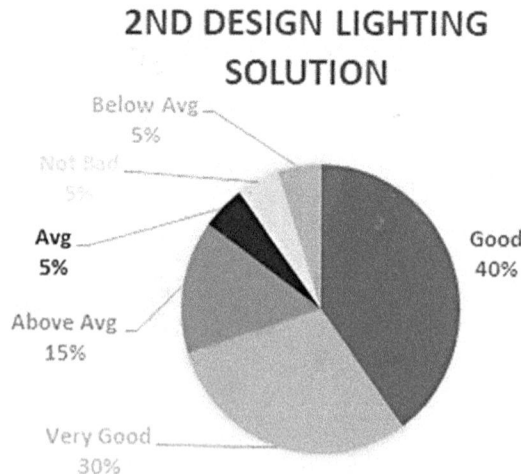

2ND DESIGN LIGHTING SOLUTION

Below Avg — 5%
Not Bad — 5%
Avg — 5%
Above Avg — 15%
Very Good — 30%
Good — 40%

FIGURE 4.19 Feedback of 2nd design lighting solution.

2ND DESIGN BUTTON FEEDBACK

No Response 5%

Bad 5%

Very Good 5%

Below Avg 14%

Good 33%

Above Avg 19%

Avg 19%

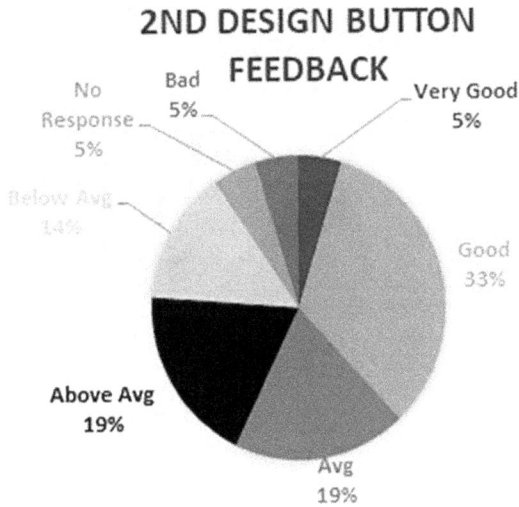

FIGURE 4.20 Feedback of 2nd design button used.

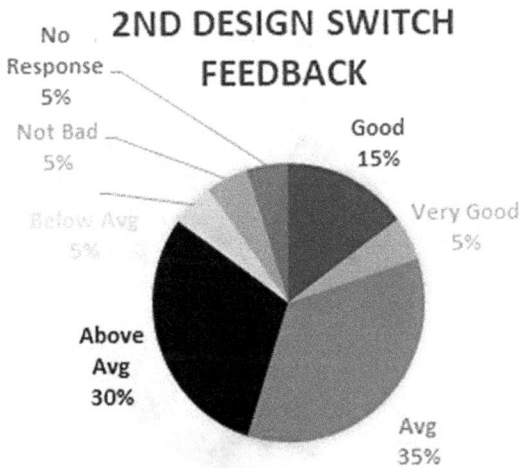

2ND DESIGN SWITCH FEEDBACK

No Response 5%

Not Bad 5%

Below Avg 5%

Good 15%

Very Good 5%

Above Avg 30%

Avg 35%

FIGURE 4.21 Feedback of 2nd design switch.

Thus, they had to be apprised on the benefits of using such a walking stick during emergent situations.

Among the participants of the feedback survey, 86% of the elderly users rendered a positive response to the handle of the third design walking stick. The handle was found to be sturdy and strong to hold the weight of the person leaning over it during a walk. Further, the handle is also ergonomically acceptable for the elderly requirements. The responses are plotted as shown in Figure 4.23.

Only 18% of the users gave a positive response to the base provided in the walking stick. This tri-junction base was found to be inconvenient for use as it creates

2ND DESIGN EMERGENCY SOLUTION

Not Bad
5%

Avg
10%

Below Avg
5%

Above
Avg
20%

Very
Good
30%

Good
30%

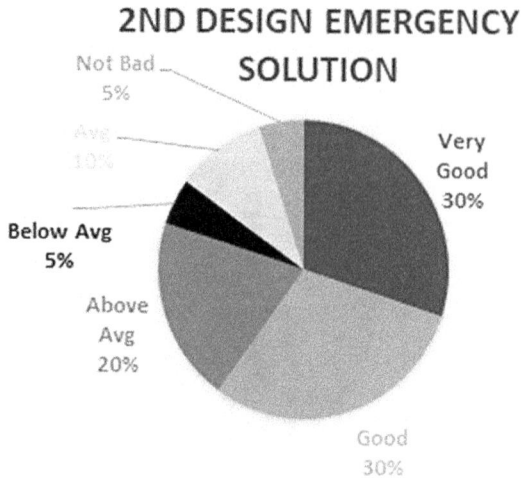

FIGURE 4.22   Feedback of 2nd design emergency solution.

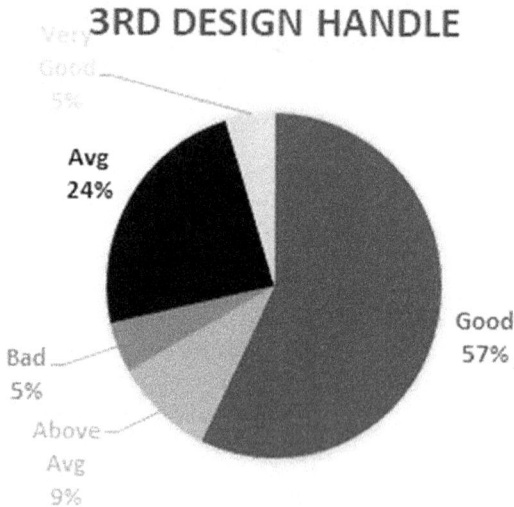

3RD DESIGN HANDLE

Very
Good
5%

Avg
24%

Bad
5%

Above
Avg
9%

Good
57%

FIGURE 4.23   Feedback of 3rd design handle.

hindrances while walking. It was suggested by most of the users to replace it with a conventional tri-base junction so that the ground contact is also firm, and it assists in walking with a strong support. Thus, the final prototype is replaced with a stronger base grip with tri-junction, based on the suggestions provided and is shown in Figure 4.7. The responses as recorded during the interview are plotted on the pie diagram and shown in Figure 4.24.

3RD DESIGN BASE

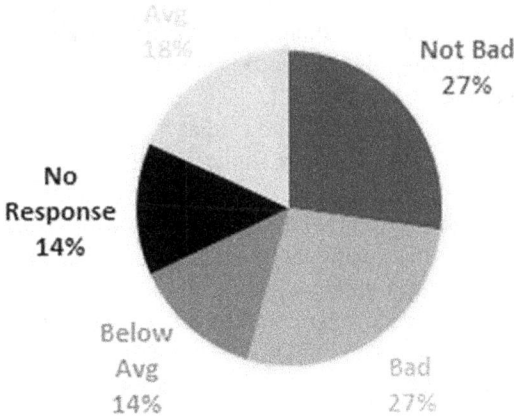

FIGURE 4.24   Feedback of 3rd design base.

Eighty-five per cent of users gave positive response to the emergency alarm and SMS solution of walking stick as proposed in the fabricated prototype of the design concept 2. 35% of the users find it to be very encouraging. The 35% users who voted in favour of the solution provided have kept themselves updated with the technological advances taking place. They are also users of smartphones and, thus, can understand the necessity of sending an SMS along with GPS location during emergent situations. The various recorded responses are plotted in pie chart and shown in Figure 4.25. The 5% of the users who voted it to be below average are those who

3RD DESIGN SWITCH FEEDBACK

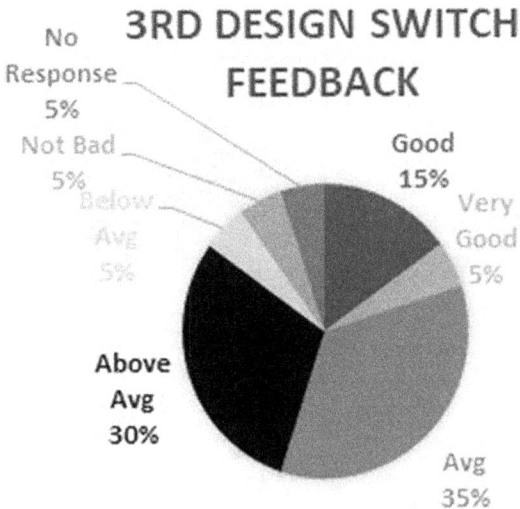

FIGURE 4.25   Feedback of 3rd design switch.

either do not have mobile phones or are still using conventional keypad phones. Thus, they had to be apprised of the benefits of using such a walking stick during emergent situations. This is almost a similar situation for the switch of Design Concept 2.

As many as 76% of the total participants in the feedback user study were of the opinion that the button to turn on the electric circuit in the walking stick, is comfortable to use. They gave positive response to the button provided in walking stick. This is plotted in Figure 4.26, which also shows that 5% of the participants could not respond appropriately. This is almost a similar situation for the switch of Design Concept 2.

A similar response like design concept 2 is found for the lighting solution for design concept 3 also. Ninety per cent of users gave positive response to lighting solution provided in the third design of walking stick. Keeping the same attributes and the reasons assessed for the response, the data is plotted on the pie diagram and shown in Figure 4.27.

Interestingly, 90% of the elderly users favoured the necessity of an emergency alarm and SMS solution provided in walking stick design concept 3. Since this design is more suited to an aristocrat kind of design, the emergency solutions were found to be a retrofit for the prototypes. The pie diagram of the data collected in the feedback session is shown in Figure 4.28.

In all, 21 secondary users were interviewed to seek feedback on the emergency solutions, as provided in the walking sticks. It was of utmost encouragement to observe that all of them gave positive responses to emergency alarm and SMS solution provided in the walking stick. The secondary users can be from any age group as they belong to the most vital contacts in the registered mobile phone of the elderly. Programming or coding can be done so that only one or multiple secondary users can

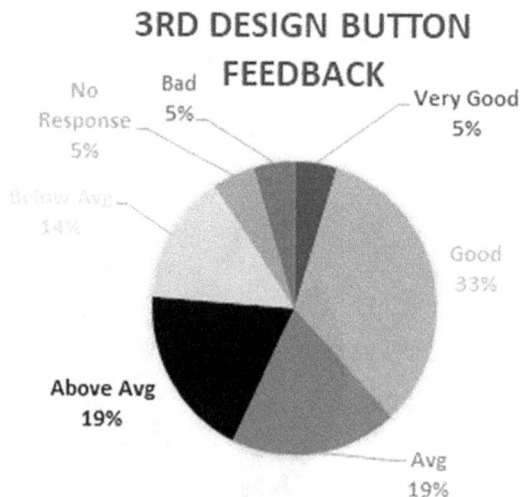

FIGURE 4.26   Feedback of 3rd design button.

3RD DESIGN LIGHTING SOLUTION

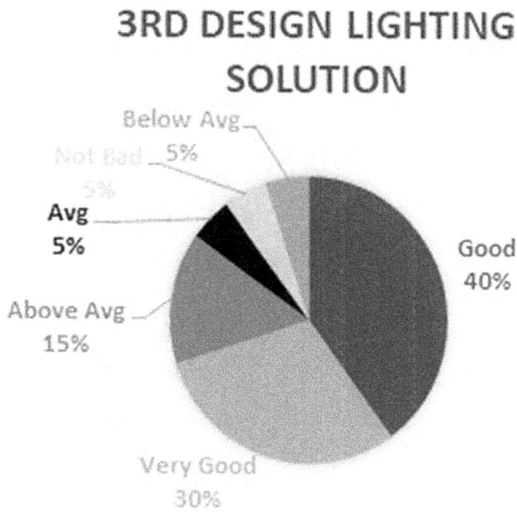

FIGURE 4.27 Feedback of 3rd design lighting solution.

3RD DESIGN EMERGENCY SOLUTION

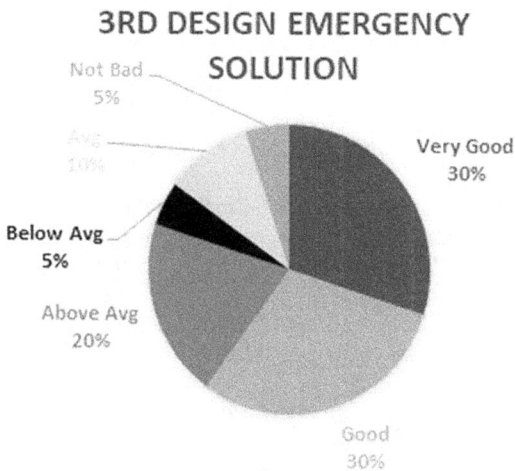

FIGURE 4.28 Feedback of 3rd design emergency solution.

receive the SMS and the GPS location of the elderly user. The pie diagram plot is shown in Figure 4.29, clearly highlighting that the communication received in this regard by the secondary users is extremely vital.

Secondary Users Feedback

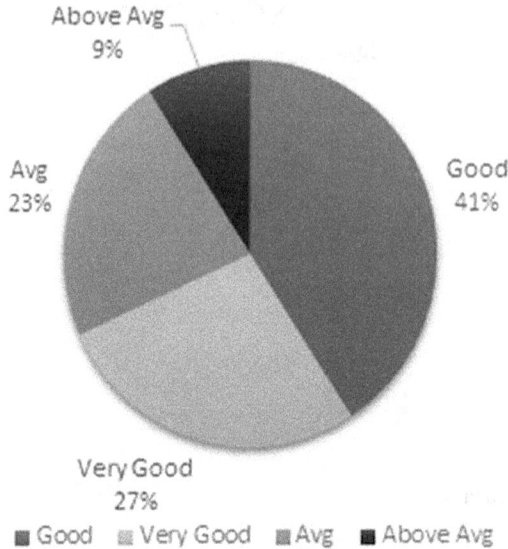

Above Avg
9%

Avg
23%

Good
41%

Very Good
27%

■ Good   ▧ Very Good   ▨ Avg   ■ Above Avg

FIGURE 4.29   Feedback from the secondary users the of 2nd and 3rd designs for the emergency solutions provided.

REFERENCES

1. Leslie Cabarga, *The Designer's Guide to Color Combinations*, Adams Media Publications, 1999.
2. Sandu Cultural Media, *Color and Space*, Gingko Press, 2012.
3. Eddie Opara, John Cantwell, *Color Works: Best Practices for Graphic Designers, An Essential Guide to Understanding and Applying Color Design*, Rockport Publishers, Beverley, MA, 2014.
4. Irima Boom, *Colour Based on Nature*, Thomas Eyck, Oosternijkerk, Netherlands, 2012.
5. Ulrich Bachmann, *Colour and Light: Materials for a Theory of Colour and Light*, Niggli, Sulgen, Switzerland, 2011.
6. Faber Birren, *Color Psychology and Color Therapy*, The Citadel Press, Secaucus, NJ, 1978.
7. Nigel Cross, *Design Thinking: Understanding How Designers Think and Work*, Berg Publishers, 2011.
8. Tim Brown, Clayton M. Christensen, Indra Nooyi, Vijay Govindarajan, "On Design Thinking", *Harvard Business Review*, 2020.
9. Bill Burnett, Dave Evans, *Designing Your Life: How to Build a Well-lived, Joyful Life*, Knopf, 2016.
10. Michael Lewrick, Patrick Link, Larry Leifer, *The Design Thinking Playbook: Mindful Digital Transformation of Teams, Products, Services Businesses and Ecosystems*, Wiley, 2018.
11. J. Robert Rossman, Mathew D. Duerden, B. Joseph Pine II, *Design Experiences*, Columbia Business School Publishing, 2019.
12. https://store.arduino.cc/usa/arduino-uno-rev3

13. https://www.adafruit.com/product/2542
14. https://invensense.tdk.com/products/motion-tracking/6-axis/mpu-6050/
15. https://lastminuteengineers.com/sim800l-gsm-module-arduino-tutorial/
16. https://lastminuteengineers.com/neo6m-gps-arduino-tutorial/
17. N. Raghu, V. N. Trupthi, N. Krishnamurthy, K. Rasagnya, D. A. Darshan, "Effects of Electromagnetic Field on Patients with Implanted Pacemakers", *International Journal of Advance Research and Innovative Ideas in Education*, Vol. 1, No. 5, pp. 38–41, 2016.

5 Re-design Issues of the Umbrella for the Elderly

Umbrellas are products that help save one from either the scorching heat in broad daylight or from rain during the monsoon or a shower. It is also an important lifeline product for the elderly in India. Many elderly people use an umbrella also as a walking assisting product. Thus, umbrellas find it designed and prototyped into many different shapes and sizes depending on the usage. For example, small umbrellas which can be folded and kept in a bag are very common with people attending offices. However, for the elderly, a conventional umbrella serves both purposes, that is walking assistance and shelter from sun or rain.

Power failure during monsoons or the rainy season is very common in many parts of the world. Also, travelling with an umbrella at night during monsoons may lead to several other possible accidents as well. Some designs and umbrella prototypes are also proposed and tested with technological intervention [1–4]. Though so, the efficacy of the structure and base with respect to average Indian elderly anthropometry and requirements remains a niche area of investigation. This chapter lays a detailed study of the pedagogy followed by are-look into umbrella design with respect to the average Indian elderly.

5.1 ANTHROPOMETRY OF THE INDIAN ELDERLY

Following the same practice as explained in Section 3.1, the relevant anthropometric details of the Indian elderly male and female are collected. The reference to the collected data is made from [5]. The standard procedure of [5] is followed for measurements and collection of the data. The statistical treatment of the data gathered is the same as followed in Section 3.1.1.

5.1.1 ANTHROPOMETRY PARAMETERS AND DATA

The data on four different dimensional measurements, presented herein, are collected from males and females of 60 years of age and above (with a ratio of 1.179: 1) from mixed occupational groups who are engaged in a variety of activities at home, offices, educational organizations, agriculture, business, industries, etc. The participants are located or co-located in different locations in India. There are a total of 85 elderly participants (46 males and 39 females).

Locations covered during the study mainly include the following areas:

1. Jabalpur (Madhya Pradesh)
2. Kadapa (Andhra Pradesh)
3. Ongole (Andhra Pradesh)

DOI: 10.1201/9781003414957-5

Though so, measurements are taken at other places in the country also; however, maximum data sets are covered from the three locations as specified. Four parameters are measured from the elderly, which are used to re-design the umbrella. These are mentioned as follows:

1. Upper arm
2. Forearm
3. Chest depth
4. Bi-deltoid

Upper Arm: The anthropometry of the upper arm is a set of measurements of the shape of the upper arms. The key measurement under this consideration is taken for the upper arm length. This is measured with a non-stretchable measuring tape. The other parameters like triceps skin fold (TSF) and the (mid-) upper arm circumference ((M) UAC) are not used for the umbrella re-design context. The triceps skin fold or TSF is the width of a fold of skin taken over the triceps muscle. It is measured using skinfold caliper, which incorporates the data of the skin fat also. The mid-upper arm circumference is the circumference of the upper arm at that same midpoint, measured with a non-stretchable tape measure. Thus, only the upper arm measurement suffices for the purpose of the design problem statement.

Forearm: The anthropometry of the forearm is a set of measurements of the shape of the forearm length. This is also measured using a simple non-stretchable measuring tape. Together the forearm and the upper arm are crucial for deciding the length of the umbrella.

Chest depth: This is the maximum horizontal distance from the vertical reference plane to the front of the chest in men or breast in women. There is a chest depth caliper which is used for measuring the chest depth of an individual.

Bi-deltoid: This is the maximum horizontal distance across the shoulders, with breadth measured to the protrusions of the deltoid muscles. The bi-deltoid and the chest depth taken together help in finding the approximate width of the umbrella.

All the protocols followed and the limitations of the measurements are the same as mentioned in Section 3.1. Interestingly some data like the grip diameter, etc., are already measured in the walking stick case; however, the same is of equal importance even in the context of handle design for the umbrella. These will be discussed during the concept generation or ideation phase of the umbrella re-design.

Table 5.1 provides a comprehensive view of the anthropometric data collected. The table provides the data pertaining to the minimum and the maximum value of the anthropometric data, the percentile from 5th to the 95th (as is the usual practice), the Mean of the data collected and the Standard Deviation. The tabular data contains the first Standard Deviation for each data set. All the measurements are strictly in a centimetre scale.

This data is collected from various cities in India and specifically from elderly Indian citizens belonging to the age group of equal to or more than 60 years of age.

Due ethical clearance was taken before collecting the data and consent from every individual was taken to collect the data.

In Table 5.1 Standard Deviation is calculated using Excel by using formula STDEV.P, which is used for calculating standard deviation of population. The calculation of the percentile is done using Equation (3.10).

The graphs in Figures 5.1 to 5.12 are also plotted following the normal distribution, based on the data collected to observe the mean and the Standard Deviations. This is done individually for males and females and is combined for the different anthropometric parameters. All the figures are plotted using MATLAB. The figures from the data collected can be plotted in any figure plotting tool based on the convenience of the users. The figures also provide a detailed input of the second Standard Deviation so as to predict any mismatch in the data collected, which may prevent the propagation of the errors in the anthropometric data so collected the design solutions provided based on them. All the measurements are in centimetres.

Upper arm male: For the upper arm measurements in Males, the calculated details are as follows:

Mean = 28.61, SD = 2.706, Mean – SD = 25.904, Mean + SD = 31.316, Mean – 2*SD = 23.198, Mean + 2*SD = 34.022. The mean, SD and all other relevant data are plotted in Figure 5.1. The figure clearly shows that a normal distribution pattern is followed, which means that the measurements were precise.

Upper arm female: The various statistical details of the upper arm measurements for females is calculated as follows:

Mean = 27.91, SD = 2.76, Mean – SD = 25.15, Mean + SD = 30.76, Mean – 2*SD = 22.39, Mean + 2*SD = 33.52. The variations and the plot are shown in Figure 5.2.

TABLE 5.1
Statistical Representation of Data Collected

S.No	Parameters		Min	Percentiles					Max	Mean	±SD
				5th	25th	50th	75th	95th			
1	**Upper Arm**	Male	20	23.625	28	29	30	32	35	28.61	2.706
		Female	21	23	26.5	28	29	32.1	34	27.91	2.76
		Combined	20	23.1	27	29	30	32	35	28.29	2.74
2	**Forearm**	Male	38	41	43	44	45	47	48	43.91	2.05
		Female	35	36	39	40	41.5	43.05	43.5	39.91	2.04
		Combined	35	37.6	40	42	44	47	48	42.07	2.86
3	**Chest Depth**	Male	11	16	18.625	20.25	21.875	27.875	29.5	20.69	3.54
		Female	15	17	20	23	25.5	29	34	23.06	4.12
		Combined	11	16	19	21	23.25	28.5	34	21.52	3.9
4	**Bi-deltoid/**	Male	32.5	35	38	40.5	43	47.75	49	40.92	3.92
	Max.	Female	27	37	35	37	38	41	43.5	36.73	2.94
	Shoulder	Combined	27	34	36	38	41.5	46.8	49	39	4.07
	width										

Min– Minimum, Max– Maximum, Mean– Simple Average, SD– Standard Deviation

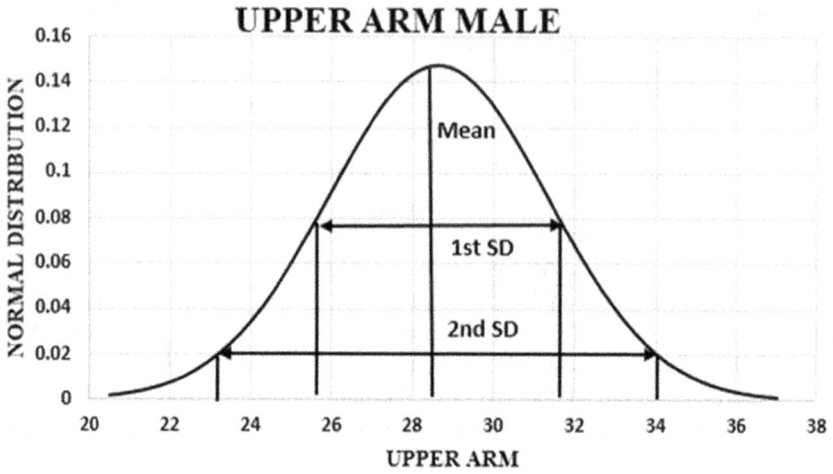

FIGURE 5.1   Graph of upper arm male data using mean and standard deviation (SD).

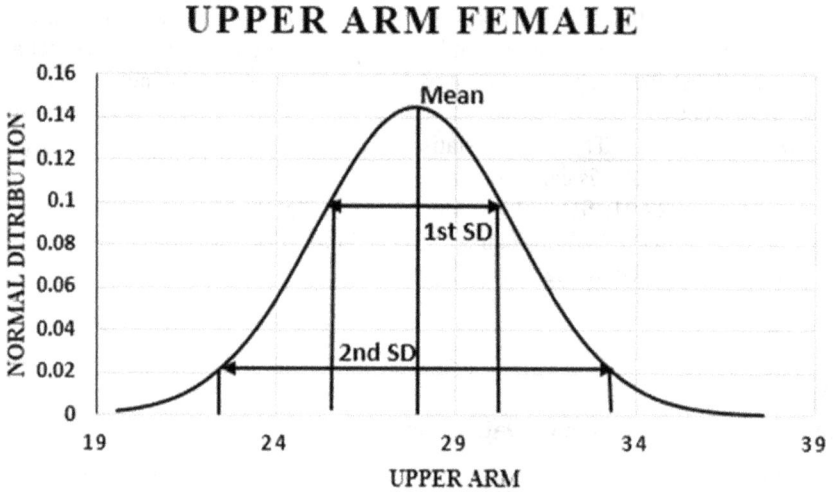

FIGURE 5.2   Graph of upper arm female data using mean and standard deviation (SD).

Upper arm combined: The various calculated details for the combined data of males and females taken together are given as follows:

Mean = 28.29, SD = 2.74, Mean – SD = 25.55, Mean + SD = 31.03, Mean – 2*SD = 22.81, Mean + 2*SD = 33.77. The plot of the data is shown in Figure 5.3.

Forearm male: The calculated parameters of statistical variations are as follows:

Mean = 43.91, SD = 2.05, Mean – SD = 41.86, Mean + SD = 45.96, Mean – 2*SD = 39.81, Mean + 2*SD = 48.01. These variations are plotted in Figure 5.4.

UPPER ARM COMBINED

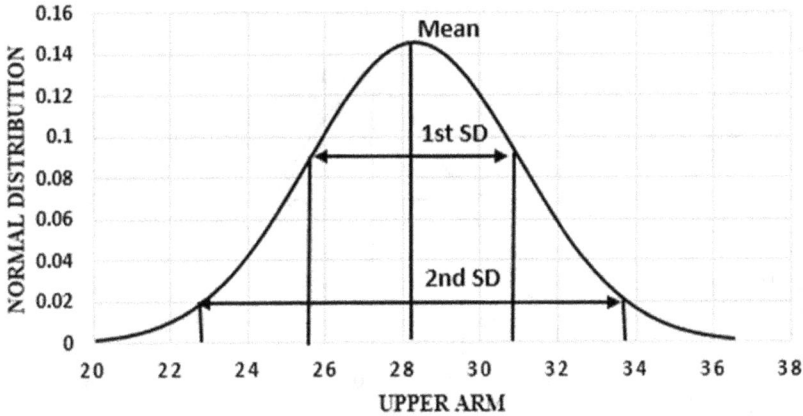

FIGURE 5.3 Graph of upper arm combined data using mean and standard deviation (SD).

FOREARM MALE

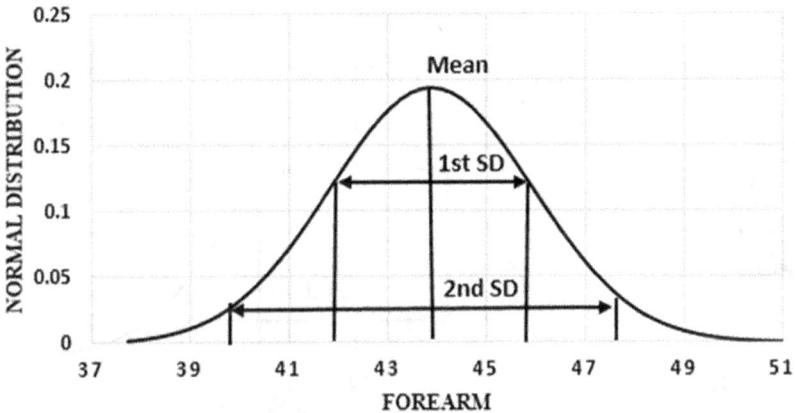

FIGURE 5.4 Graph of forearm measurements of male data using mean and standard deviation (SD).

Forearm female: The statistically treated data for the forearm measurements of female elderly is as follows:

Mean = 39.91, SD = 2.04, Mean − SD = 37.87, Mean + SD = 41.95, Mean − 2*SD = 35.83, Mean + 2*SD = 43.99. The plot showing the variation of the data is shown in Figure 5.5.

Forearm combined: The statistical data for the forearm measurements of the combined form, that is, both males and females taken together are as follows:

Mean = 42.07, SD = 2.86, Mean − SD = 39.21, Mean + SD = 44.93, Mean − 2*SD = 36.35, Mean + 2*SD = 47.79. The plot of the data is shown in Figure 5.6.

FOREARM FEMALE

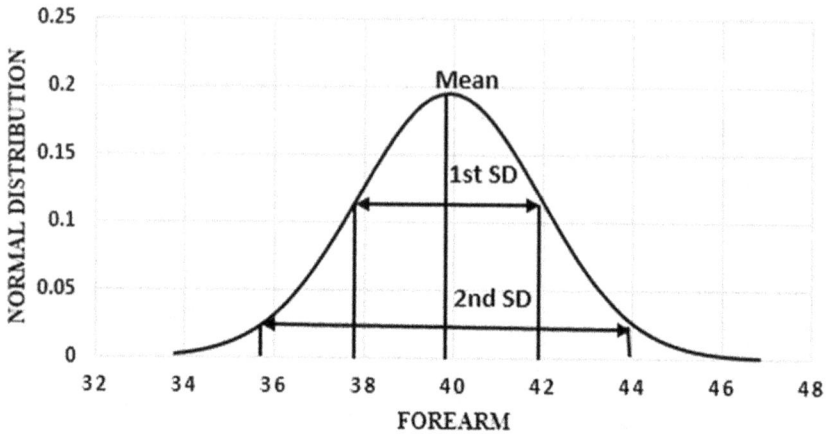

FIGURE 5.5   Graph of forearm female data using mean and standard deviation (SD).

FOREARM COMBINED

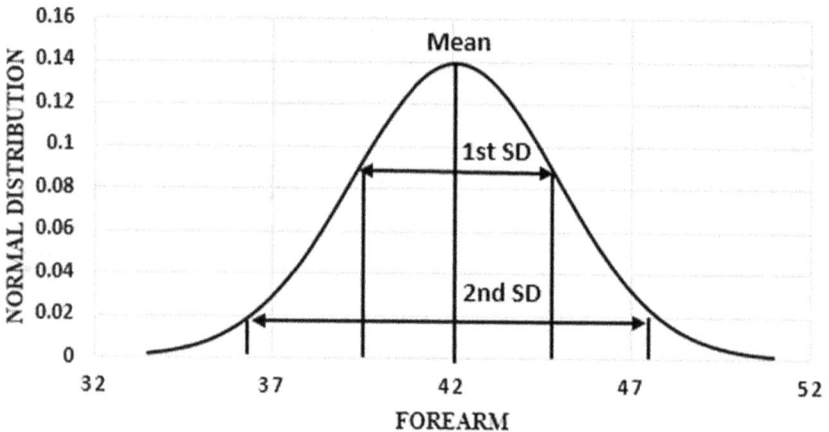

FIGURE 5.6   Graph of forearm combined data using mean and standard deviation (SD).

Chest depth male: The calculated data which is statistically treated is presented below:

Mean = 20.69, SD = 3.54, Mean – SD = 17.15, Mean + SD = 24.23, Mean – 2*SD = 13.63, Mean + 2*SD = 27.77. This is shown in the plot of Figure 5.7.

Chest depth female: The calculated data which is statistically treated is presented below:

Mean = 23.06, SD = 4.12, Mean – SD = 18.94, Mean + SD = 27.18, Mean – 2*SD = 14.82, Mean + 2*SD = 31.3. This is shown in the plot of Figure 5.8.

CHEST DEPTH MALE

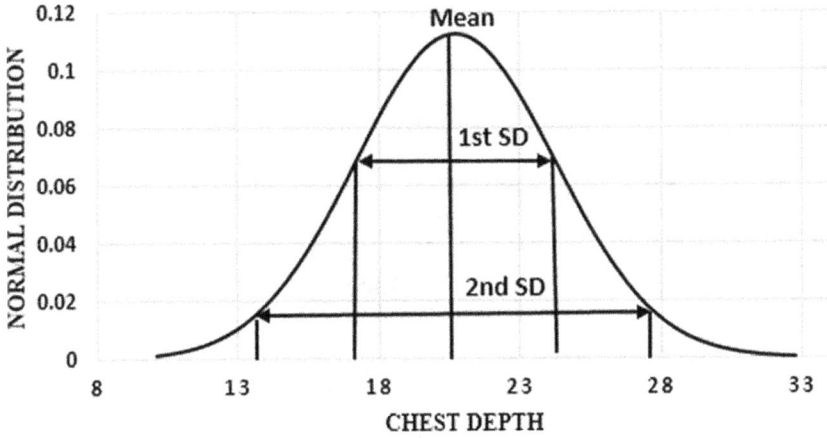

FIGURE 5.7 Graph of chest depth male data using mean and standard deviation (SD).

CHEST DEPTH FEMALE

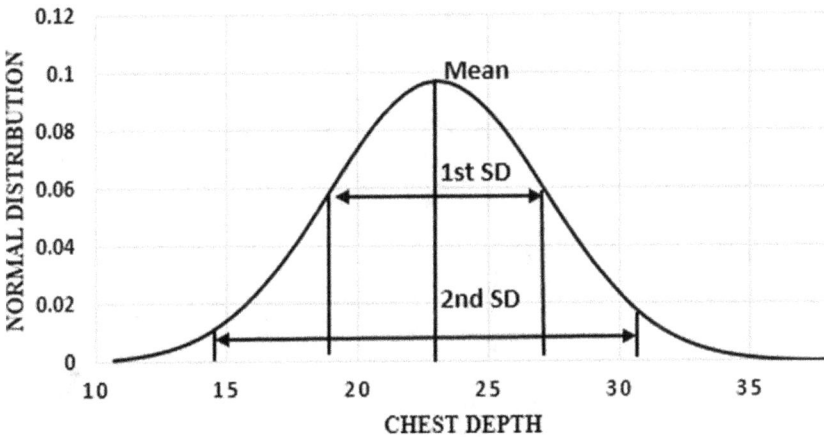

FIGURE 5.8 Graph of chest depth female data using mean and standard deviation (SD).

Chest depth combined: The statistical data for the chest depth measurements of the combined form, that is, both males and females taken together are as follows:

Mean = 21.52, SD = 3.9, Mean – SD = 17.62, Mean + SD = 25.42, Mean – 2*SD = 13.72, Mean + 2*SD = 29.32. This variation is plotted in Figure 5.9.

Max. shoulder width male: The calculated data which is statistically treated is presented below:

Mean = 40.92, SD = 3.92, Mean – SD = 37, Mean + SD = 44.84, Mean – 2*SD = 33.08, Mean + 2*SD = 48.76. This is shown in the plot of Figure 5.10.

CHEST DEPTH COMBINED

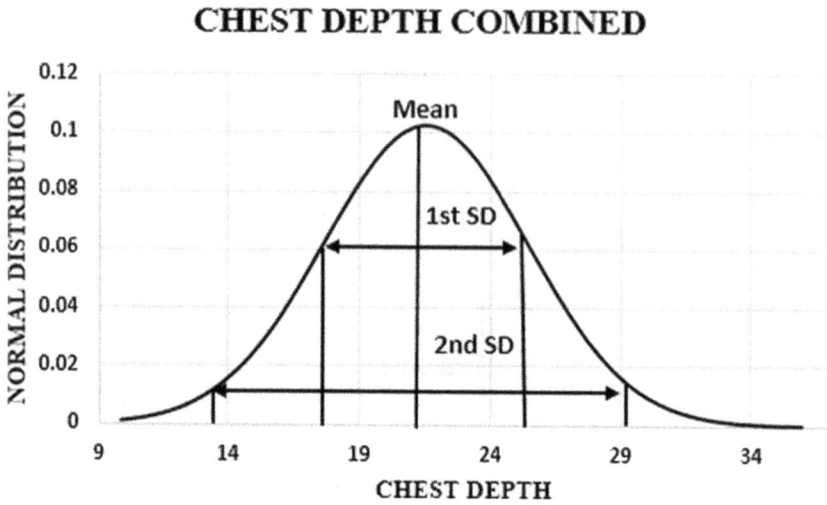

FIGURE 5.9   Graph of chest depth combined data using mean and standard deviation (SD).

MAX. SHOULDER WIDTH MALE

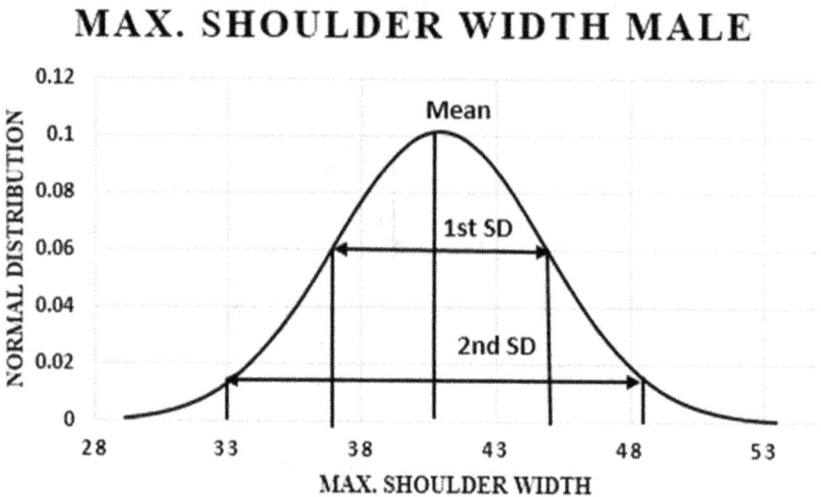

FIGURE 5.10   Graph of maximum shoulder width for the male elderly data using mean and standard deviation (SD).

Max. shoulder width female: The calculated data which is statistically treated is presented below:

Mean = 36.73, SD = 2.94, Mean − SD = 33.79, Mean + SD = 39.67, Mean − 2*SD = 30.85, Mean + 2*SD = 42.61. This is shown in the plot of Figure 5.11.

Max. shoulder width female: The statistical data for the maximum shoulder width measurements of the combined form, that is, both males and females taken together are as follows:

MAX. SHOULDER WIDTH FEMALE

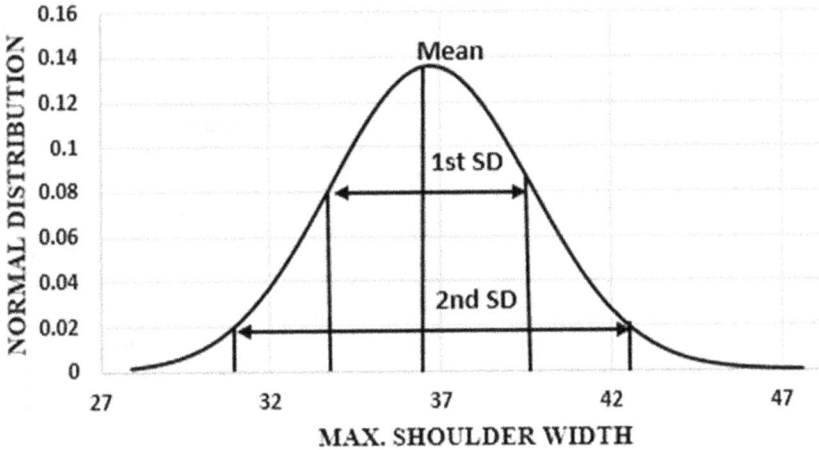

FIGURE 5.11 Graph of maximum shoulder width for the Female elderly data using Mean and Standard Deviation (SD).

Mean = 39, SD = 4.07, Mean − SD = 34.93, Mean + SD = 43.07, Mean − 2*SD = 30.86, Mean + 2*SD = 47.14. This is shown in the plot of Figure 5.12.

If a data distribution is approximately normal, then about 68% of the data values are within one *standard deviation* of the mean (mathematically, Mean±SD, where

MAX. SHOULDER WIDTH COMBINED

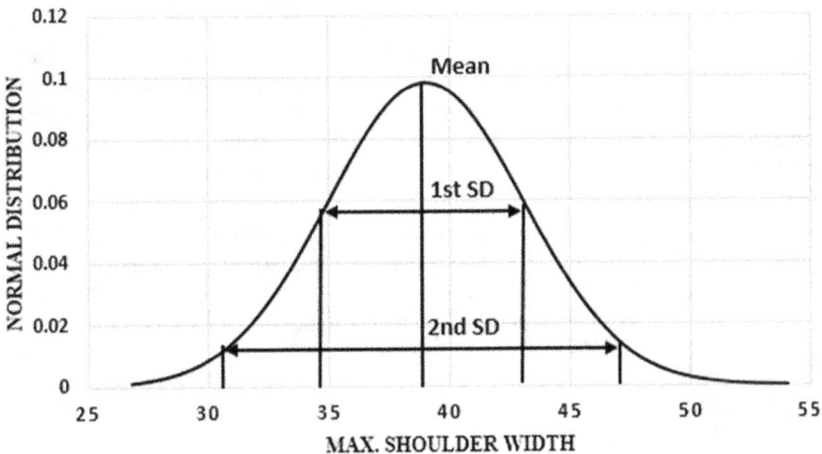

FIGURE 5.12 Graph of maximum shoulder width of the combined data for both male and female elderly taken together using mean and standard deviation (SD).

mean is arithmetic mean), about 95% are within two *standard deviations* (Mean±2*SD) and about 99.7% are within three standard deviations (Mean±3*SD). The graphs consist of only one and two standard deviations of all anthropometry references.

5.2   DESIGN OF THE QUESTIONNAIRE FOR DIRECT OBSERVATION AND ANALYSIS

For understanding the use of the umbrella and its necessity in daily life of the elderly, the direct observation and activity analysis with interview and questionnaire process is adopted. This is the same as the process adopted and explained in Section 3.2 for the Walking Stick. Following the same process, a sample size of 27 participants (all elderly males and females above the age group of 60 years) provides the input to the questionnaire. All concepts of design thinking, also to avoid the Hawthorne effect, are taken into consideration during the interview process [6–8]. The questionnaire designed for interviews included both qualitative and quantitative questions. Certain Qualitative questions for seeking opinions on Likert scale have also been chosen for umbrella uses. Also, the Likert Scale of 1 to 7 is also discussed with its reason for the choice of such a scale.

The questionnaire consists of both Qualitative and Quantitative questions in nature. The questions along with relevant explanations are clearly marked with each question. The qualitative questions are highlighted in bold and italics. The quantitative questions are highlighted in bold and underlined as well. There are some notes added shown as Bold plus italic plus underline. This helps in understanding the elderly user better.

The questionnaire is discussed below:

Age: ....**(It is a Quantitative question which is asked towards the end of the interview session to avoid the Hawthorne effect. If this question is posed at the beginning of the interview itself, the likelihood that the responses received by the elderly shall be biased is likely to be very high.**)

Gender: ....**(It is a Quantitative question which is marked or used for data analytics and ideation after the interview session of all the interviewees is completed. Sometimes the preferences of the male and female choices differ based on Gender and so it constitutes an important point in the questionnaire.)**

1.   What type of umbrella do you use? (**It is a Quantitative question which is to be plotted on the pie chart graph. The users are shown two options as in Figure 5.13. Figure 5.13a is the conventional manual umbrella and Figure 5.13b is the foldable umbrella which can be kept and stored conveniently. The aim of this question is to find out whether the elderly user finds it comfortable to use the conventional umbrella or foldable umbrella, irrespective of the varied designs which are available in the market for these two types.)**

2.   What type of grip does it have? (**This is a Quantitative question which shall be plotted on the pie chart graph. Through this question, the different possible handles available in the market popularly shall be shown and**

the choice made by the users shall be recorded. In all, four variations of the handle are identified. This is shown in Figure 5.14. Figure 5.14a is a conventional U-shaped handle, Figure 5.14b is a cylindrical handle, Figure 5.14c is an automatic push button-type handleand Figure 5.14d is a cylindrical handle with grooves for a better grip to hold the umbrella.)

3. How many times do you use the umbrella in a day? Is it specific to any season only? *(This is a Quantitative question which shall be analysed and plotted on the pie chart. The reasons can be qualitative; however, common reasons can be grouped together and then analysed quantitatively also.)*

4. Do you encounter any pain/discomfort in your palm while holding the umbrella? If yes, mark the area. *(This is a Qualitative question. This is used for plotting the pain points on the palm picture as shown in Figure 5.15. The pain points can be on both sides of the palm while holding the umbrella, i.e., the dorsal side as shown in Figure 5.15a or the palm side as shown in Figure 5.15b.)*

5. On a scale of 1–7, how much pain/discomfort do you feel? *(This is a Quantitative question mapping the qualitative pain point area on the Likert scale.)*

    (Least) 1---2---3---4---5---6---7(most)

    Likert Scale: Quantitative analysis of the Qualitative data; user maps the rate of discomfort. Scale of 1–5 is not chosen since it is not clear that some people might have injuries and so it might lead to inherent pain which cannot be mapped on ascale of 1–5. Scale of 1–10 is not chosen since this scale is chosen for more complex scales,and since this is a focused group on the elderly, scale of 1–10 is avoided. Thus, a medium region comprising of a scale of 1–7 is adopted.

6. Do you encounter any pain/discomfort in your arm while using umbrella for a long time? If yes, mark the area. *(This is a Qualitative Question where the user marks area of discomfort in the arm region. Since the weight of the*

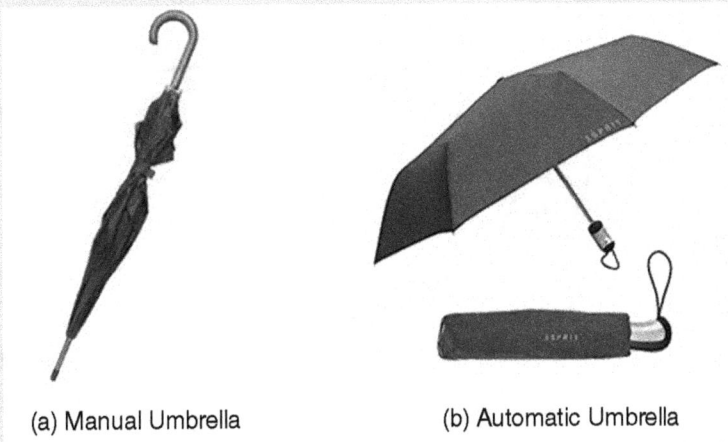

(a) Manual Umbrella   (b) Automatic Umbrella

FIGURE 5.13 Different types of umbrellas available in the market. (a) The conventional umbrella (b) Folding umbrella.

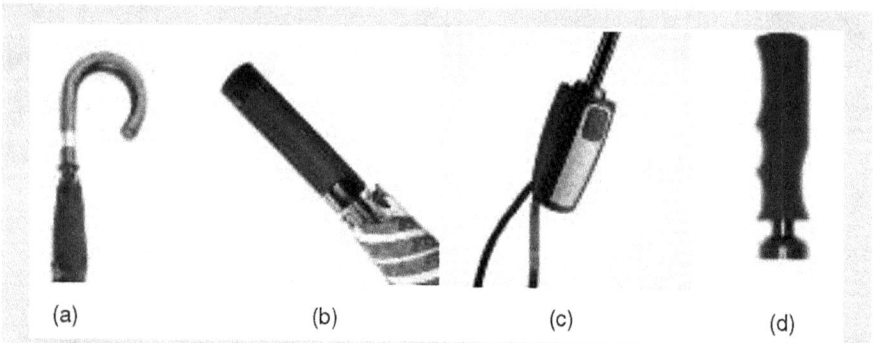

FIGURE 5.14 Different types of handles of umbrellas. (a) Conventional U-shaped handle (b) Cylindrical handle (c) Automatic push button type handle (d) Cylindrical handle with grooves.

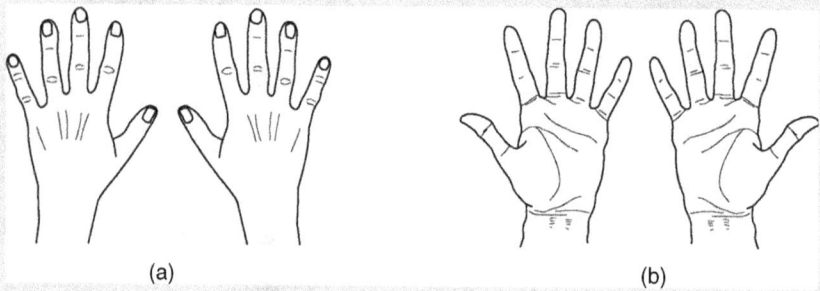

FIGURE 5.15 Discomfort mapping on the hand (a) Dorsal side of hand (b) Palm side of hand.

FIGURE 5.16 Discomfort or pain mapping on the forearm and upper arm.

umbrella lies on the shoulder through the arms, the marking of the pain points on it is extremely crucial.)

7. Where do you store the umbrella when you are?

    At home: _____

    Outside: _____

    (This is a Qualitative question. The aim of posing this query is for making compact and easy design. There are specifically two different locations where the umbrella can be used:while we use the umbrella while going out from home or while we are already outside home. Both these cases are sought as inputs from the elderly users.)

8. Do you face any difficulty while locating it? *(This is a Qualitative question which asked for finding a design solution for issues related to locating the product. Is it handy enough so that the elderly can easily find it and then use it? This is the main motto behind the question.)*

9. What kind of problems do you face while using the umbrella? *(This is a Qualitative question. The rationale behind this is to understand the interaction with the user and any issues which can be solved during design ideation stage or by technological intervention.)*

10. Do you face any difficulty while opening and closing the umbrella? *(This is also a Qualitative question. The aim behind this is to understand the interaction with the user while operating the umbrella. The operation mechanism of the manual and the automatic umbrella are different. Through this question, this operation mechanism is observed, and the problems associated are articulated.)*

11. How do you drain umbrella after use in rains? *(This is a Qualitative question. Usually during the monsoons, the umbrellas are also left open for drying purpose. This question intends to ask how the elderly drains the umbrella so that it can be reused again.)*

12. Which hand do you hold the umbrella majority of the time? *(This is a Qualitative question to understand the working hand of the elderly for using the umbrella. There may be cases when the elderly might be a right-handed person but may prefer to use the umbrella on the left shoulder or vice versa.)*

13. Do you have any suggestions on how it could be improved? Would it be beneficial if the umbrella had a torch/light source? Location tracker? *(This is a Qualitative question. The aim of this question is for identification of technical and design implementations to cater to the elderly user needs. The suggestions given by the elderly users shall be taken into consideration for the ideation stage and the prototyping purpose.)*

5.3 DESIGN THINKING APPROACH USING DOUBLE DIAMOND AND EMPATHY MAP

After the Anthropometric data is available and the Questionnaire is designed, the double diamond map approach of Design Thinking is used, as explained in Section 3.3. While implementing the double diamond map for umbrella study, the definitions for the different section and empathy mapping of the data collected are discussed in detail as follows.

1. *Discover (Research):* The design process according to the double diamond model begins with the discovery of user needs, current design and choice of various methods for the same. This is in tandem with Figure 1.6 of Section 1.3. Some of the important patents as applicable in umbrella design are discussed below, which are referred from the references [9–14]:

   Illuminated umbrella: This product refers to an invention that relates to an illuminated umbrella intended to assure Safety of the user of the umbrella in rainy weather, comprising a light Source in a handle or a Shaft, thereby illuminating at least part of the Shaft coupled to the handle, and therefore the inside of the umbrella, and the face of the user or the vicinity are illuminated by lighting the shaft itself which is near the face. This is used in order to notice the presence of a human or the user emphatically to others and drivers, so that the Safety of the user in traffic may be enhanced.

   Umbrella: The umbrella comprises of a hollow cylinder or central tubular member which serves as the handle for the umbrella and as a storage chamber for the assembly of the parts comprising the top cover; a plurality of flexible stays for supporting the cover in open position. A similar number of tension chords control and brace the supporting stays, and sliding operating members in the tube, to which these parts are connected, and by means of which they are operated in opening and closing the umbrella.

   Umbrella and mount assembly for wheelchair: A shaft and mounting assembly for an umbrella for an umbrella suitable for use on a wheelchair are proposed. The assembly is designed to permit a wheelchair bound person to install, deploy and retract the umbrella. It is also adapted to permit unobstructed Viewing to the front, hand-free operation and a working area to be left to the patron when the umbrella is deployed. It is also designed to permit navigation through portals adapted for wheelchair access. It may also be adapted to facilitate stability in wind and rain.

   Multifunctional umbrella: The utility model relates to a multifunctional umbrella added with an entertainment function, which comprises an umbrella fabric and an umbrella handle, wherein the umbrella handle is embedded with an MP3 player and a light-change LED bulb which are connected in parallel, a button battery and a solar battery supply power for the MP3 player and the light-change LED bulb, respectively. The button battery is embedded in the umbrella handle and the solar battery is covered on the upper surface of the top part of the umbrella fabric. The button battery and the solar battery are connected in parallel, and the light-change LED bulb is controlled independently by a switch arranged on the umbrella handle.

As the multifunctional umbrella is provided with an MP3 player and a light-change LED bulb, a user can listen to music while walking and simultaneously can enjoy continuously changing colourful lights, thus effectively alleviating umbrella-holding fatigue, and making travel interesting.

Based on the above inputs of the various umbrellas present, the methodology for the research remains the same. It includes Shadowing, interviews, Questionnaires, Qualitative and Quantitative data, Likert scaling. To avoid errors, Mom's test, rechecking and testing for bias is done.

2. *Define (Insights):* This refers to defining the areas of interest and development by analysis of data collected in the previous stage. Factors associated include Storage, Reason for use, Parallel tasks and Illumination. This is explained as follows:

Storage: Storage of the umbrella for the primary user, that is, the elderly user of the umbrella in different contexts of use and settings is to be identified. This gives an insight into the varying settings where aid can be used and how its use changes from one setting or context to the other.

Parallel tasks: This gives an insight into how elderly user behaviour alters from usual when they are given an umbrella and helps understand the extent of impact it can cause.

Illumination: To be visible the conventional umbrella is basically black finished for being readily seen by others in the vicinity. However, it is only effective in the daytime and may be inconspicuous or invisible at night to threaten the security of the aged person in walking.

The consolidated response of all the 27 users interviewed is tabulated as shown in Table 5.2.

In the above table, some elderly did not want to explicitly mention their age, so very politely, their age group was found, and they responded as over 60 years of age. This is mentioned as '60+' in Table 5.2.

3. *Analysis:* Once the responses are extensively brain stormed and detailed to avoid missing any critical information while providing the design solutions, the designer steps closer to the ideation concept. However, among the first step for the analysis is mapping the pain points while using their usual umbrellas.

Users were asked various parts of their palm and arm where they felt pain while using their umbrella. The elderly users then marked various spots on the image of their palm and arm. Some of them also showed the areas on their own palms and arms for a clear elucidation. The areas of the arm and the palm shown by them were then plotted on the image attached. These images were then traced and converted into 10% opacity overlays and then all marked on one common hand frame to pinpoint areas which were common in the complaint about pain. The higher the opacity of the area, the usual the spot was for pain or for discomfort. This is graphically explained below in Figure 5.17a for the palm discomfort and Figure 5.17b for the arm discomfort.

TABLE 5.2
User Study of Walking Stick

S.No	Gender	Age	Type of Umbrella	Type of Handle	Storage at Home	Storage at Outside
1	Male	60+	b	c	Cupboard	Bag
2	Male	81	a	a	Cupboard	Hanging to shoulder
3	Male	76	a	a	Hanging on wall	Hanging to Shoulder
4	Male	60+	b	c	Hanging on wall	Hanging to wrist
5	Male	60	a	a	Almirah	Hanging to the hand
6	Male	71	a	a	Hanging on wall	Hanging to the hand
7	Male	60	a	a	Hanging on wall	Hanging to the shoulder/hand
8	Male	79	a	a	Hanging on wall	Hanging to the hand/fixing in between the hand and the body
9	Male	65	a	a	Hanging on wall	Hanging to the shoulder
10	Male	68	b	c	Hanging on wall	Hanging to the shoulder
11	Female	60	b	None	Hanging on wall	Keeping in hand
12	Female	63	b	None	Cupboard	Keeping in hand
13	Male	60+	b	c	Hanging on wall	Hanging to hand
14	Female	73	b	c	Cupboard	Bag
15	Male	60+	a	a	Hanging on wall	Keeping in hand
16	Female	60+	b	b	Cupboard	Keeping in hand
17	Female	60+	a&b	a&c	Hanging on wall	Keeping in hand/Bag
18	Male	62	b	c	Hanging on wall	Keeping in hand
19	Female	71	b	c	Hanging on wall	Hanging to hand
20	Female	60+	b	c	Hanging on wall	Hanging to hand
21	Male	60+	b	c	Hanging on wall	Hanging to hand
22	Female	60+	b	c	Hanging on wall	Keeping in hand
23	Female	60+	a	a	Hanging on wall	Keeping in hand
24	Female	60+	a	a	Hanging on wall	Keeping in hand
25	Female	60+	a	a	Hanging on wall	Keeping to hand
26	Female	60+	a	a	Hanging on wall	Keeping in hand
27	Male	75	b	c	Hanging on wall	Bag

Once the pain points are mapped, the chunking of the data is done. This step included identifying the quantum of users that fall under different categories depending on their usage and preference. The users were asked 'what they did with the umbrella when not in use for short and long duration of time (for storage)'. A repetition was observed in the responses after interviewing a less than 25% of the participants, and thus the data was separated into the major categories and plotted on a pie chart to show the similarity in user preference and behaviour.

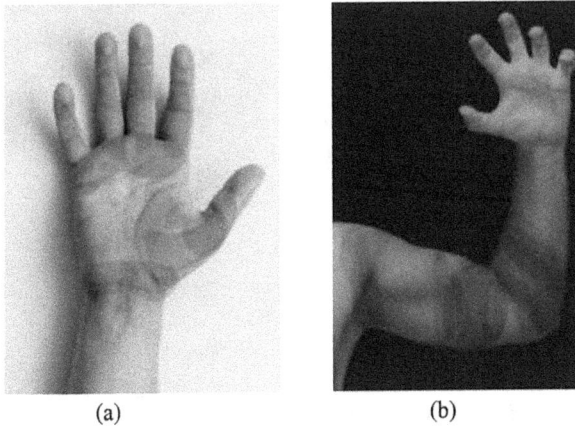

(a) (b)

FIGURE 5.17 Pain and discomfort points on (a) Palm (b) Arm.

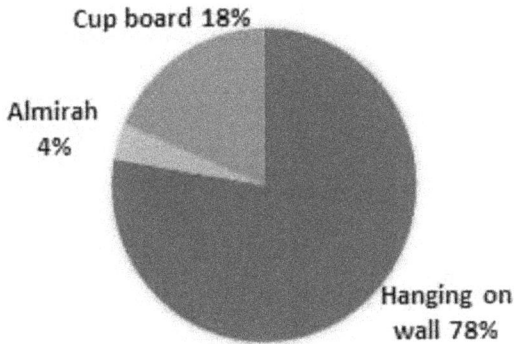

FIGURE 5.18 Graphical data of storage at home.

From the plotted data, it was found that cupboards and hanging on the wall are the most preferred locations of storage for longer durations of umbrella and for short duration users preferred to keep the umbrella in hand and hang the umbrella from the hand. From the chunked data it can be observed that for storage at home, from Figure 5.18, people generally hang their umbrella from the wall at a convenient place. This is generally close to the bedside or near the exit of the main entrance of the house. Seventy-eight per cent of the people responded as hanging on the wall.

In Figure 5.19, it is shown that 38% of the people keep an umbrella in the hand while travelling with it. Those who favour the folding or the automatic umbrella prefer to keep the umbrella in their bag. This constitutes around 14% of the participants.

Fifty-four per cent of elderly people prefer an automated umbrella as referred from Figure 5.20, while 46% prefer manual umbrella. Figure 5.21 shows that 46% elderly prefer an umbrella with a conventional U-shaped handle, while 42% elderly people prefer a single-push button automated

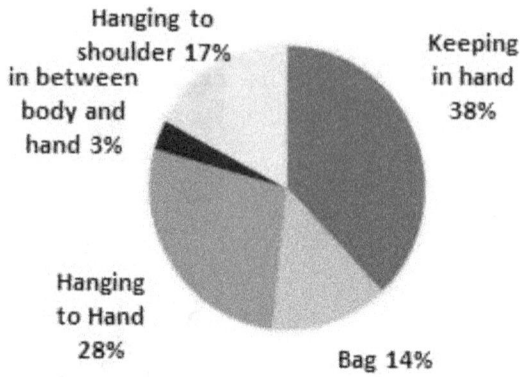

FIGURE 5.19 Graphical data of storage at outside.

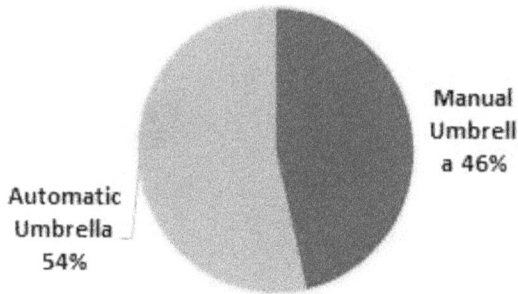

FIGURE 5.20 Graphical data of umbrella (Refer Figure 5.13).

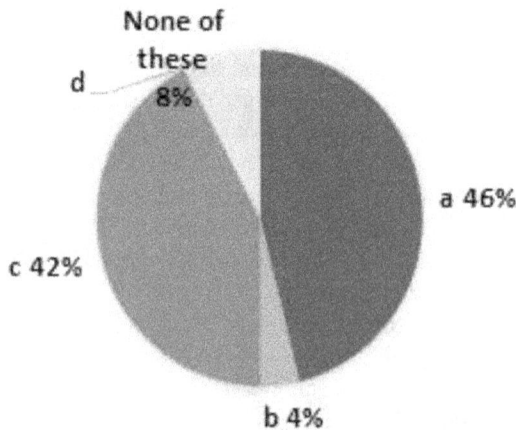

FIGURE 5.21 Graphical data of handle (Refer Figure 5.14).

umbrella handle for regular usage. The reference to the handles is taken from Figure 5.14, which was shown to the elderly users who participated in the survey.

The inputs of the chunked data pave the necessity to understand the requirements of an umbrella from the elderly users. The pain points and the qualitative questions help identify the need of a lighting solution integrated with the umbrella, which is a dire need while an elderly travel at night even for short distances during monsoons. The insight reveals that the main purpose of umbrella for elderly is to protect from sun or rain or sometimes also use it for balance while walking. Also, the weight of the umbrella and a good grip to reduce pain points need to be introspected while proposing the design solutions apart from technological intervention.

4. *Empathy maps:* After the chunking of the data is done, the information is further placed in the empathy map to understand the main requirements and modifications that need to be done while re-designing the umbrella. The empathy map for the 27 users collectively is placed in Figure 5.22. The various quadrants of the empathy map are explained below:

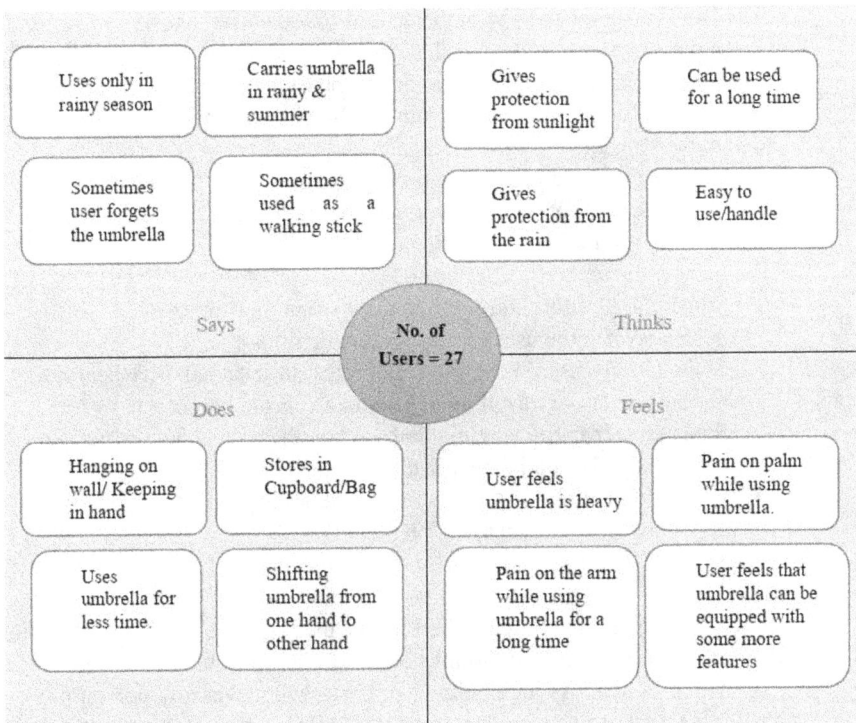

FIGURE 5.22 Empathy maps for the elderly users of umbrella study.

Says
- Says quadrant contains what the users say loud in an interview or user study.
- Here in the empathy map, when the elderly users were asked about usage of umbrella, some users said that they carry an umbrella during rainy and summer seasons and some users said that they use an umbrella only during the rainy season.

Thinks
- Thinks quadrant captures what the user is thinking throughout the user experience.
- Based on observations of users, conclusions of thinks quadrant are drawn.
- While using the walking stick,
  1. Some users think that an umbrella might give protection from the sunlight and rainfall.
  2. Some users think that an umbrella can be used for long time.
  3. Some users think that it is easy to use/handle the umbrella even during a walk. This can be used to provide stability as well while walking.

Does
- The Does quadrant deals with the actions that the user takes.
- From the user study, some conclusions are drawn for the Does quadrant.
- Through the interview, it is found that some of the elderly users have a tendency of shifting the umbrella from one hand to another hand while using due to the pain and discomfort in palm and arm.
- Also, some elderly users use umbrella for a less time due to the weight of the umbrella or the pressure it exerts on the shoulder or the arm or the hand, while using it for a longer duration of time.
- For storing the umbrella after usage,
  1. Some elderly users store it in the bag when they are outside. This is possible when the user uses an automatic umbrella.
  2. Some elderly users keep their umbrellas in their hands, when they are outside or are travelling or going away from home.
  3. Some elderly users hang the umbrella on their shoulders, when they carry it outside their home. But this may be for a short duration of time.
  4. Some elderly users stated that they store or keep the umbrella in their cupboard or leave it hanging on the wall.

Feels
- The Feels quadrant reflects the elderly user's emotional state of mind set. This gives some important details on how the users are connected to their products. As an example, in India, whenever any person buys a vehicle whether two wheelers or four wheelers, the first thing that people do is to pray for the car through a priest. Similarly, Indian elderly are connected to the products of their daily use which is marked in the Feels quadrant.

- While using the Umbrella
    1. Some users feel umbrella is heavy to use.
    2. Some users feel pain/numbness on palm when the umbrella is used for a long time.
    3. Some users feel pain/discomfort in upper arm and forearm when used for long time.
    4. Some users feel that umbrella is lacking of features for comfort ability and emergency.

Based on all the quadrants of the empathy map, the ideation process takes place. The 'develop' stage of the double diamond map reflects the stage where all ideations are presented. This is explained in the next section, that is, Section 5.4 Design Concepts. Further, 'deliver' stage reflects the fabricated prototypes of the ideated concepts. This is discussed in detail in Chapter 6.

5.4 DESIGN CONCEPTS

Various designs are worked with different feasible solutions, following the Indian anthropometric dimensions of the elderly and with the following constraints in mind:

1. The proposed ideated design solutions of the umbrella should be in tandem with the Indian elderly's mental model.
2. The ideation solution should be low-cost and easy-to-install products with reference to the existing umbrellas (as elderly hold great sentimental value for old products).
3. The ideation should help in reducing the static load of using the umbrella in case of long-term usage.
4. The ideated design solution should provide an ergonomic solution to the handle grip of the umbrella.
5. A lighting solution or illumination to the pathway should be provided in the umbrella. As is repeatedly reported by the elderly users that they face extensive trouble in using the umbrella during nights or monsoons or during power cuts. Thus, lighting solution need to be incorporated in the ideated solution. It has been observed as a practice that the elderly users carry a torch separately along with the umbrella when they travel outside their home during night (in case of monsoons) or during any power outage.
6. The ideated solution must facilitate the movement of handle to restore original or traditional usage as well.

Based on these inputs and the discussion carried above using design thinking approach, two different concepts are proposed which are deliberated as follows:

Concept 1: This concept is based on reducing the pain points on palm and arm by using an Asymmetric umbrella design which is tied up near the shoulder so as to transfer the weight of the umbrella near the shoulder junction. Since the shoulder junction acts as a truss, the weight is divided and the user feels practically no weight in carrying the asymmetric umbrella.

The key features of the umbrella are stated below:

A. It offers a better grip of the handle.
B. The concept tries to remove the pain or discomfort that exists due to long-term use and weight or pressure executed on the shoulder or arm or palm regions.
C. A handcuff is also provided to fix the umbrella to the hand near shoulder, which gives further support to the elderly user and reduces the strain on palm.
D. Switch provided near the handle is used for illumination and an emergency alarm is also provided for the elderly user in times of emergency.

Figure 5.23 shows design concept 1 of the asymmetric umbrella. The anthropometry reference to concept 1 is tabulated in Table 5.3.

One of the key features of design concept 1 is that the umbrella cover and cloth is asymmetric in nature. Thus, the cross-section of the open umbrella is not circular as is true for conventional umbrellas; it is rather elliptic in nature. This helps to cover the person completely as the umbrella is tied to the edge of one of the arms.

FIGURE 5.23   The ideated asymmetric umbrella concept 1.

TABLE 5.3
Anthropometry Reference to Figure 5.23

Figure 5.23	Anthropometry	Dimension
Handle width	Handbreadth without the thumb, at the metacarpal	11 cm
Grip diameter	Grip inside diameter	4.2 cm
Length	Forearm & Upper Arm	100 cm
Diameter	Max. Shoulder width	100 cm

FIGURE 5.24 The ideated Umbrella Concept 2.

Concept 2: The walking stick concept 2 is an extension of the existing automated umbrellas with space provided for retrofitting the electronics to support the technology intervention. The grip is kept cylindrical so as to match
the human grip. Also, the grip area shall have a switch and hollow section to
place the electronic circuitry.

The handle length is 10cm, which includes 90-mm palm length of 95%
elderly male and extra 10mm for hand dynamicity. The handle diameter is
4.2cm, which is to cater to the 25% of the elderly female inner grip diameter enabling better power grip to all. The proposed design is as shown in
Figure 5.24.

5.5   DISCUSSION AND CONCLUSION

The study follows the design thinking methodology for creation of ideas that solve
the issues with the current designs of umbrella to the elderly, specifically in the
Indian context. The comparison of ideas and current designs in the market on the

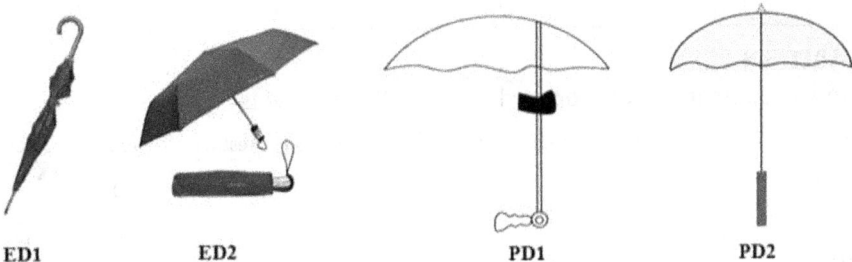

ED1   ED2   PD1   PD2

FIGURE 5.25   Existing designs (ED) compared to proposed designs (PD).

basis of various qualities is also presented. Figure 5.25 shows existing and proposed designs of the umbrellas, which are then compared based on minimized Likert scale.

The ED 1 (Existing Design 1) shows the conventional manual umbrella with a standard U-shaped handle. The ED 2 (Existing Design 2) is an automated umbrella which can be easily folded and kept or stored in a bag while travelling. PD 1 (Proposed Design 1) is the picture of the proposed Asymmetric Umbrella as an outcome of the ideation of the design Concept 1. The PD 2 is the picture of the proposed umbrella with an extended arm grip and better ergonomic handle as an outcome of the ideation of design concept 2.

All the designs were taken on a common platform for evaluation of the proposed concepts with the existing solutions.

Table 5.4 compares the Proposed Designs with Existing Designs. The features of the designs are evaluated on a 0–2 scale (minimized Likert scale). The various features used for the comparison includes Balance, Grip, Ergonomic Features, Storage issues, Illumination, Aesthetics and Support System. The balance feature assesses whether the umbrella is strong enough that it can provide adequate balance to the elderly user while using it in all conditions of day and night. Further, the support features adjudge whether the presented umbrella is capable of lying independently without holding it physically. The support feature is present only in the PD 1, since a wristband can easily hold the umbrella, even when the elderly user is not holding the umbrella with his or her hands separately.

The definitions of the scores are explained below:

0 – No presence of the feature
1 – Weak presence of the feature
2 – Strong presence of the feature

It is interesting to note that the proposed designs scored more score than Existing designs based on the minimized Likert scale.

The detailed study through Direct Observation and Activity Analysis helped propose two design concepts of Umbrella for the Indian elderly. The first concept helps reduce pain in the palm and arms. The second helps integrate technology by providing relevant space for housing the electronic circuitry. The subsequent chapter deals with the detailed fabrication procedure of the two concepts proposed.

TABLE 5.4

Design Comparison of Proposed with Existing Products

Designs	Balance	Grip	Ergonomic	Storage	Illumination	Aesthetic	Support	Total
E.D 1	2	0	1	2	0	1	0	6
E.D 2	2	1	2	2	0	1	0	8
P.D 1	2	2	2	1	2	1	2	12
P.D 2	2	2	2	2	2	1	0	11

E.D. – Existing Design, P.D. – Proposed Design

REFERENCES

1. "Parts of an umbrella" at the Wayback Machine, Carver Umbrellas, 2007.
2. Yiyang Cai, Chunyang Zhang, Nichen Niu, "An Intelligent Umbrella Design Scheme based on Node MCU", *International Journal of Engineering and Applied Sciences*, Vol. 5, No. 11, pp. 7–9, 2018.
3. Koray Korkmaz, "Generation of a New Type of Architectural Umbrella", *International Journal of Space Structures*, Vol. 20, No. 1, pp. 35–42, 2005.
4. Meher Dev Gudela, Atharv Kulkarni, Abhishek Dhotre, Kshitiz Srimali, "Design of an Automatic Umbrella Actuated through Water and Temperature Sensors", *Journal of Mechatronics and Robotics*, Vol. 4, No. 1, pp. 191–210, 2020.
5. Debkumar Chakrabarti, *Indian Anthropometric Dimensions for Ergonomic Design Practice*, National Institute of Design, 1997.
6. Siddhu Kulbir Singh, *Methodology of Research in Education*, Sterling Publisher, New Delhi, 1992.
7. S. P. Sukhia, P. V. Mehrotra, *Elements of Educational Research*, Allied Publisher Private Limited, New Delhi, 1983.
8. Martyn Denscombe, *The Good Research Guide*, Viva Books Private Limited, New Delhi, 1999.
9. Akira Tatsumi, "Illuminated Umbrella", US Patent number 5848831 dated December 15, 1998.
10. Daniel J. Thode, "Illuminated Umbrella", US Patent number US2007/0189002A1, dated August 16, 2007.
11. Thomas Szumlic, Jan Gallagher, "Umbrella and mount Assembly for Wheelchair", US Patent number US2004/0103934A1 dated June 3, 2004.
12. Edward Allee, "Universal Wheelchair umbrella and shealth", US patent number US 6378539B1 dated April 30, 2002.
13. Wang Quiling, Xu Dong, Li Jing, "Multifunctional Umbrella", Chinese patent number CN2011200169350U dated August 10, 2011.
14. Xu Weimin, "Multi-functional safety umbrella", Chinese patent number CN87211089 dated July 13, 1998.

6 Technology Intervention in the Designs of the Umbrella

The previous chapter dealt with the details of the design thinking approach to propose two design concepts. These design concepts are based on the anthropometric collected from among 87 Indian elderly users (males and females included). After the anthropometry data is collected, the direct observation and activity analysis approach is adopted to observe the uses of the umbrella and how the umbrella is adopted as a product by Indian elderly users. Based on the inputs received, graphical chunking of the collected data is done, and then the empathy map is also plotted. Based on this detailed exercise, two design concepts are also proposed for the umbrella redesigned in the Indian context. These design concepts are then fabricated to build prototypes and then tested among users to collect feedback of the umbrellas so developed. The user testing is a field testing where the inputs are taken from the elderly again using Direct Observation and Activity Analysis.

Two designs of the umbrellas were fabricated. Material selection, concept of design, electronic circuits implementation, physical limitations of the user and psychological interpretations of the user were kept in mind while fabricating. The various prerequisites for the fabrication purpose are stated below:

- Mechanical Stability – It is important to observe that the umbrella fabricated should be mechanically stable. This is especially of concern keeping in mind the asymmetric umbrella which has no symmetricity in its canopy structure. Thus, the mechanical stability of structure to withstand any air drag or heavy rains becomes a cause of concern.
- Lightweight – The fabricated structure should be light in weight. If it becomes heavy to carry for any reason, be it the mechanical constraints or the electronic circuitry, then it again defeats the purpose of re-designing the umbrella.
- Non-conducting material – The material for the umbrella should be non-conducting. Since it will house electronic circuits with a battery connection and recharging capabilities, it is essential to note that there should be no short circuit path that affects the elderly user while using the electronic part of the umbrella. Fatal shocks, if any, can only invite trouble for the elderly using the product.
- Space to house electronic circuit – It is interesting to observe that an umbrella has an extremely small space to house the battery or any electric or electronic circuit. This brings a limitation on the space to be identified in an umbrella to suitably host the complete circuit. Further, since the umbrellas are used for protection from rain or monsoons, the protection of the electronic circuit from moisture becomes extremely vital to observe.

DOI: 10.1201/9781003414957-6

6.1   TECHNOLOGICAL INTERVENTION THROUGH PROCESS FLOW

The design thinking process also has a very important tool for analysis known as the journey map [1–5]. A journey map is a visualisation of the process that a person goes through in order to accomplish a goal. The journey map is an effective tool to compile the series of user actions into a timeline. Next, the timeline is populated with user thoughts and emotions in order to create a narrative. This narrative is condensed and polished, ultimately leading to visualisation. The narratives in the sections of the journey map are responsible to provide the apt solution needed for the problem statement. This similar process is also adopted in the case of technological intervention of the walking stick re-design process with various feasible solutions.

The detailed concept of a journey map is discussed in Section 4.2. The adaption of the process flow from the journey map for the fabrication and technology intervention of umbrellas is shown in Figure 6.1. There are some stages in order to accomplish the goal, which is technological intervention. The various stages are presented in Figure 6.1.

The different stages of the process flow are described below:

I.   Define
   • Defining different types of umbrellas that are present in the market is the main motto of this stage.
   • This helps to identify which type of umbrellas most people use.
   • The various problems that users face while using the umbrella are articulated in this section. The answer to the question of why there is a problem is also laid out in this stage.
   • This stage also tries to identify the technological solution to the problem, that is, technological incubation in the umbrella for the comfort of the elderly.
   • At this level, the concept of lighting solution to be provided to elderly users in times of darkness and at the time of power outages using Light Emitting Diode (LED) strips as one of the possible solutions is also explored.
   • Emergency alarms and sending location details to the registered number are some of the other solutions in times of emergency that are defined at this stage. The details of the technological advancements shall be explored as the stages of the journey map proceeded.

II. Enquire
   • To achieve desired technological solutions, enquiring and selecting the desired electronics is an important stage in the process.
   • There are a lot of electronic modules available in the market which gives the same kind of output. But choosing the one which occupies less space and low power consumption and offers low complexity are the prima facie to be retrofitted in the design solutions.

Scenario: Designing a Umbrella for elderly which will have technological solutions in times of user requirement and emergency.

Expectations: Electronic components required to fulfill the technological solutions.

I. Define	II. Enquire	III. Check	IV. Test	V. Compare	VI. Select	VII. Installing the electronic components in design	VIII. Testing
1. Define present umbrellas 2. Define the problems with present umbrellas 3. Define the Situations where the present umbrellas are lacking to provide the solutions. 4. Define the technological Intervened solutions to such problems.	5. Enquiring for required technological components which provide the desired output. 6. Selecting the components which provide the desired output.	7. Check whether the components fit in the space of designed umbrellas or not.	8. Test the electronic components whether they are fulfilling the requirement or not. 9. Testing the components individually and when all are connected.	10. Compare the components which provide the desired output and fit in the design, to sort out the best.	11. Select components which are best for design.	12. Fitting the components to umbrellas according to requirement.	13. Testing whether the components are working or not.

FIGURE 6.1   Process flow of technological intervention for umbrellas.

- In the case of microcontrollers, Arduino Series (Nano, Uno, Mega) and Raspberry Pi, etc., are all available for commercial and research use.
- In the case of GSM modules, SIM Series, which are 900A, 800L, 808 etc., are already available in the market.
- In the case of power supply to the circuit, 3.7-volt batteries are available of different sizes, shapes and capacities that can be fit into any electronic circuit.
- For the working of electronic modules, coding is required for interfacing the electronic components/sensors with the microcontroller selected. Earlier codes were mostly Assembly language instructions as used with 8051, etc.
- Enquiring about coding and programming language is also an important aspect.
- C++/C language coding is popularly used for deploying instructions to the microcontroller (e.g., for Arduino UNO, C/C++ are used).

III. Check
- Since the electronic components are housed inside the umbrella, they have to be checked whether they can fit in the constrained space of the designed/fabricated sticks or not. A small space to incubate the electronic circuitry along with accommodating the power supply is a major challenge.
- In the first ideated design concept, LED strips and buzzers are used for lighting and emergency alarm. GSM module is used due to the space constraint. Since this is an asymmetric umbrella, the technological intervention in this module is kept simplistic and not complicated.
- In the second ideated design concept, SIM 800L (GSM) module, LED strip and buzzer are used. This module helps communicate with the secondary users who can receive an emergency SMS about the elderly user.
- For the power supply, 3.7-volt batteries are used along with TP 4056 module, which is used for the charging of the batteries.

IV. Test
- Before incorporating circuits inside the umbrellas, components must be tested individually whether they fulfil the requirements and needs of the elderly user or not. They should also be tested whether they supply the desired output or not.
- For testing purposes, software coding can be done for modules individually. This is in tandem with the Universal Design principle of Chunking or Top-Down approach.
- Due to bugs in coding, sometimes the modules do not generate the desired output.
- While testing, if it is found that the battery connected to SIM 800 L modules must be charged separately using TP 4056 module, the battery has to be removed from the circuit and has to be charged separately. This is a major constraint in using this module.

- TP 4056 module has B+, B– icons in it. The + (positive polarity) and – (negative polarity) of the battery must be connected respectively to charge the battery.

V. Compare
- To sort out the best, the various electronic components and circuits must be compared with each other. This gives an idea of the most-apt solution leading to low cognitive load on the elderly while using the umbrella.
- It is extremely important to solve the bugs in the software coding for interfacing the microcontroller. Using trial and error method, one can sort out the code used for the program.

VI. Select
- This refers to selecting various components which are best suited for the proposed design solutions.
- It also refers to selecting a particular set of codes best suited for the circuit application.

VII. Installing electronic components in design
- After selecting the components, the components are installed into the respective walking sticks according to the requirements.
- Installing various components and connecting the components to ensure the system-level integrity is an important stage. The various modules designed are now connected in a bottom-up approach so that the system can work cohesively.

VIII. Testing
- After installing all the components, testing is necessary whether the connections made are correct or not.
- Coding for the total circuit is an important step of the testing procedure.
- Testing also includes whether the stick is working properly or not after aggregating the electronic circuitry and is an important stage in this process.

Based on the Process discussed above, the various electronic components are identified which can be retrofitted into the design to fabricate the prototypes of the umbrella with technological intervention. The different components used are as follows:

1. SPDT switch – A Single Pole Double Throw (SPDT) switch is a switch that only has a single input and can connect to and switch between two outputs. This means it has one input terminal and two output terminals.
2. SIM 800L – It is a GSM module [6]. It can connect to a global GSM network with any 2G/3G SIM.It can be operated using a microcontroller.
3. Buzzer – A buzzer or beeper is an audio signalling device, which may be mechanical, electromechanical or piezoelectric (piezo for short) in nature. The inputs can be of any mentioned form that generates a buzzer sound or alarm.

4. TP4056 battery charging module – It is used to charge 3.7-volt Li-Po battery used in the circuit [7]. A lithium polymer battery, or more correctly lithium-ion polymer battery (abbreviated as LiPo, LIP, Li-poly, lithium-poly and others), is a rechargeable battery of lithium-ion technology using a polymer electrolyte instead of a liquid electrolyte. High-conductivity semi-solid (gel) polymers form this electrolyte. These batteries provide higher specific energy than other lithium battery types and are used in applications where weight is a critical feature, such as mobile devices, radio-controlled aircraft and some electric vehicles.

5. LED strip – An LED strip is used to provide adequate luminance at night with minimal power supply from the battery or the source connected.

6. Push button – A push button switch is a small, sealed mechanism that completes an electric circuit when you press on it. When it is turned on, a small metal spring inside makes a contact with two wires, allowing the electricity to flow.

7. Arduino Pro Mini – The Arduino Pro Mini is a microcontroller board based on the Atmega328.

   It has 14 digital input/output pins (of which 6 can be used as PWM or Pulse width Modulation outputs), six analogue inputs, an onboard resonator, a reset button and holes for mounting pin headers. A six-pin header can be connected to an FTDI cable or Sparkfun breakout board to provide USB power and communication to the board. It requires coding on a platform called Arduino IDE for functioning [8]. FDTI stands for Future Technology Devices International, which is a Scottish Private company which are experts in serial bus to USB cable manufacturers.

8. Slide switches – In the proposed technological solutions provided, a three-pole two ways sliding switch is used. Slide switches are mechanical switches using a slider that moves (slides) from the open (OFF) position to the closed (ON) position.

The above components are used to provide technical inputs to the umbrella and provide appropriate technological intervention. The proposed umbrellas give lighting solutions to the elderly when dark, alarm for emergencies and SMS alert to the phone numbers registered in the code of the microcontroller deployed in the umbrella. These specific features were selected because of their relevance to elderly users based on the study carried out before. Only two features were selected because too many features would increase the cognitive load on the user, and also its circuit implementation would require more space in the umbrella, which is a stringent limitation. Also, the more the circuits added, the heavier the umbrella becomes for the user.

There are two types of solutions proposed based on the availability of space in the umbrellas. These are discussed as follows:

Lighting solution and emergency alert (Alarm and SMS): The circuit is initiated with only one trigger, which is button operated. If the button is pressed for a short duration it will trigger LED (On/Off). The button needs to be pressed for a long duration till a buzzer sounds. The buzzer will produce an alarm for 20–25 seconds for emergency help from the people in the near vicinity. After the buzzer is raised, the GSM module sends an emergency SMS to the registered mobile number/numbers depending on the kind of programming or coding burnt in the microcontroller.

The limitations of this scheme mainly are mentioned below:

First, GSM module must have signals for connecting to the network. If the elderly user is in the basement where there is no booster or in shadow regions where no signal is received on the SIM card that is inserted in the module of the microcontroller, then there is a strong likelihood that the GSM module might not work efficiently.

Second, the battery of the GSM module needs to be charged from time to time. The charging point shall be conveniently provided in the fabricated prototype, but the elderly user needs to charge the battery on a regular basis for the electronic circuit to operate.

Lighting solution and emergency alert (alarm): The circuit is controlled by the three-way SPDT switch, which is fixed to the handle of the umbrella. An SPDT switch is used to produce two outputs from one input. In this case, the switch is connected to the LED strip and buzzer. If the switch is pressed in the forward direction, LED will glow. If the switch is pressed in the backwards direction, a buzzer will produce an emergency alarm.

Two designs of umbrellas are fabricated. Material selection, concept of design, electronic circuits implementation, physical limitations of the user and psychological interpretations of the user were kept in mind while fabricating the prototypes.

The key prerequisites for the fabrication of the prototypes are mentioned below:

- *Mechanical stability* – The umbrella fabricated should have precise mechanical stability so that it can be operated conveniently and can withstand the harsh conditions of strong winds, heavy rains, etc., as is administered sometimes in the monsoon season.
- *Light weight* – If the umbrella becomes heavy, then there is a very strong likelihood that it might exert strong pressure on the shoulders of the elderly, making it extremely uncomfortable for use. The elderly, given the conditions of age factor that they bear, might not be capable enough to bear a heavy umbrella and walk along with it.
- *Non-conducting material* – The umbrella should be manufactured using non-conducting materials as it will house electronic circuits for alarm and lighting solutions and a GSM module. A battery is also to be accommodated in the design to power the electronic circuit. Thus, it becomes extremely crucial to ensure that no short circuit or a conducting path is provided which might give electric shock or cause a fatality for the elderly user.
- *Space to house electronic circuit* – The major constraint of technology incubation with the umbrella as a product is space. The umbrella does not provide much space to the designer to house a large circuit and still keep the design light in weight. The smaller circuits so designed should be accommodated strictly in the limited space of the umbrella design.

6.2 ASYMMETRIC UMBRELLA CONCEPT: FABRICATION, FEATURES AND TECHNOLOGY INTERVENTION

The Asymmetric umbrella designed is meant to shift the load of palms and the arm to the shoulder junction through a cloth strip, so as to provide ease and comfort to the elderly while using it. Thus, the technology intervention in the same is kept simplistic in nature.

The key features of the fabricated prototype are as follows:

- Materials used – Wood and metallic pipes are preferred. Wood is used near the handle where the battery and the microcontroller are placed. The components are then extended along the umbrella though connecting wires which run through the frame made of metal so as to keep the umbrella light and rugged support. This is elucidated in Figure 6.1.
- This design provides an extended length of the umbrella so that it can easily retrofit the am cloth piece and carry the weight on the arms instead of the shoulders. This is as shown in Figure 6.2 clearly.
- Handle of the umbrella is made using wood.
- Wood is carved into a cylindrical shape so as to facilitate the grip of the elderly user.

(a)   (b)

FIGURE 6.2   The prototype of the design concept 1. (a) Handle with switch and LED strip. (b) Opened umbrella with the handcuff.

- Wood is cut in a shape that will help to insert electronic components inside it. It is a small hollow section to house the complete electronic circuit.
- Electronic components used include TP 4056 module for charging, three-way switch, buzzer, LED strip and 3.7-volt battery.
- Connections of electronics:
  1. B+ (Battery) to B+ (TP 4056), −(Battery) to B−(TP 4056) is connected so that the battery can be charged conveniently.
  2. Out+(TP 4056) to LED & buzzer, Out−(TP 4056) to Switch for facilitating the required connections of charging.
  3. GND of (LED, buzzer) to switch to provide the return path to the electric current as is conventional to do in electric circuits.

The fabricated prototype is clearly shown in Figure 6.2a and b, highlighting each of the specific part clearly.

The workflow of the electronic circuit is as shown in Figure 6.3. A simple three-way switch is used to turn on the LED and raise an alarm.

The asymmetric part is achieved by adding additional joints in the frame of the umbrella, and the canopy cover is extended with additional cloth material to cover the umbrella section properly. The detailed flowchart of the actions performed based on Figure 6.3 is shown in Figure 6.4. The flowchart clearly shows the functions

Buzzer   3-way Switch   LED Strip

FIGURE 6.3   Process flow of buzzer and LED strip for the asymmetric umbrella.

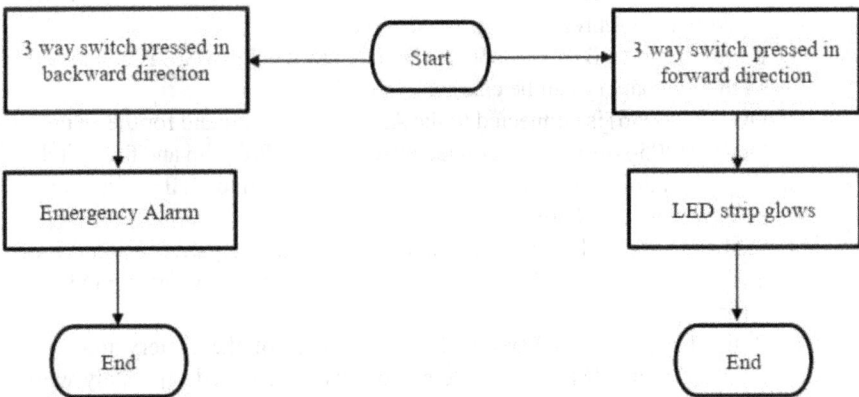

FIGURE 6.4   Flowchart of the proposed three-way switch solution as technology intervention in design concept 1 of the asymmetric umbrella.

triggered when the three-way switch is pressed in either direction. This design is kept minimalistic and so does not include any GSM module. It either raises an alarm using a buzzer or turns on the lighting solution using the LED strip.

6.3   CONVENTIONAL UMBRELLA CONCEPT: FABRICATION, FEATURES AND TECHNOLOGY INTERVENTION

Design concept 2 is aimed at introducing some rigorous electronic concepts so as to connect to the secondary users of the umbrella for the elderly. In order to achieve it, microcontroller programming facilitates both emergency SMS and lighting solutions. The second design was made using plastic handles. Electronic circuit is housed inside the handle. The key features of the umbrella are as follows:

- The material used for the handle is plastic. This is so chosen since the handle is elongated from the conventional handle for accommodating the electronic circuits inside it.
- This design is an extension of the existing compact umbrellas.
- Electronic circuits are fixed inside the handle of the umbrella. This is done by ensuring that the microcontroller casing is retrofitted in the handle section. There is a charging module also connected to charge the battery from time to time.
- Holes are drilled in the plastic casing of the umbrella connected to the handle to fix the switch and the button.
- Badminton racket grip cover is used for covering the handle. This grip provides additional capabilities to hold the umbrella properly in order to meet the requirements of the elderly.
- Electronic components that are used include TP 4056 for charging, a three-way sliding switch, buzzer, Led strip, 3.7-volt Battery, SIM 800Land Arduino Pro Mini. Two 3.7-volt batteries are required for this electronic circuit as one is used for the GSM module and the second is for LED strip.
- The various connections and interconnections of the electronic circuits deployed in the umbrella design are stated as follows:
    1. B+ (Battery) to B+ (TP 4056), −(Battery) to B−(TP 4056) is connected so that the battery can be charged conveniently
    2. Out+(TP 4056) is connected to the Arduino Pro Mini and for the switch, Out-(TP 4056) terminal is connected to Arduino Pro mini and the switch.
    3. The remaining connections from Arduino Pro Mini to all the other components connected are well maintained.
    4. SIM 800L module must be connected with one separate battery and connections of the SIM 800L module are Vcc to + (battery), GND to − (battery).
    5. If the battery is discharged, the connections of the battery must be removed with SIM 800L module and must be charged separately with TP 4056 module.

6. For charging the battery, the connections are + (battery) to B+ (TP 4056) and − (battery) to B-(TP 4056).

7. After charging the battery, it must be connected to the SIM 800L module.

Figure 6.5a shows the folded umbrella, Figure 6.5b shows the various parts of the electronic circuitry placed inside the umbrella handle and Figure 6.5c shows the LED turned ON by press of the button on the umbrella handle.

There is a possibility that the elderly might be using a pacemaker. A person with a pacemaker can use the umbrella that is designed without GSM modules. The first design of the asymmetric umbrella does not have electromagnetic components that can affect the pacemaker. But while using the second Design model a user with pacemaker has to keep the handle of umbrella at least 10cm away from the pacemaker [9] because the pacemaker is affected by the electromagnetic waves. The second Design can be used by pacemaker, but the user has to hold the umbrella with their right hand. Then only the distance from the pacemaker to the handle of umbrella will be more than 10cm.

The workflow of the electronic circuitry used in the umbrella prototype is shown in Figure 6.6. The figure shows that on a short press, the LED strip is turned ON and switched OFF in the same way. For a longer press, the microcontroller (Arduino Pro Mini) is activated to send the SMS (Short Message Service) to the contacts of the elderly. For implementing the workflow in C/C++ codes, the flowchart is given in

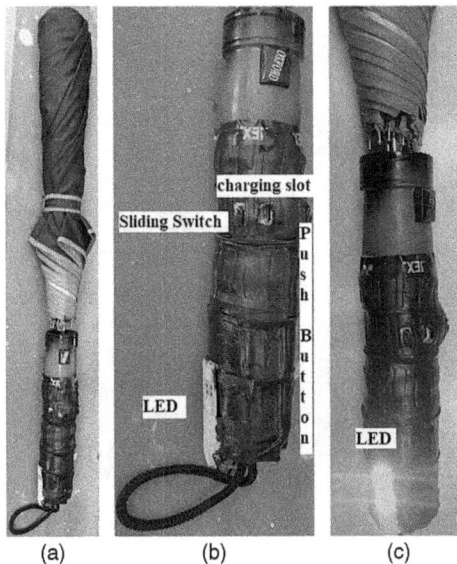

(a)   (b)   (c)

FIGURE 6.5   The fabricated prototype of design concept 2 of the umbrella for the elderly. (a) Folded umbrella (b) Handle modified by the insertion of the electronic circuitry in the handle (c) The glowing LED strip for providing lighting solution.

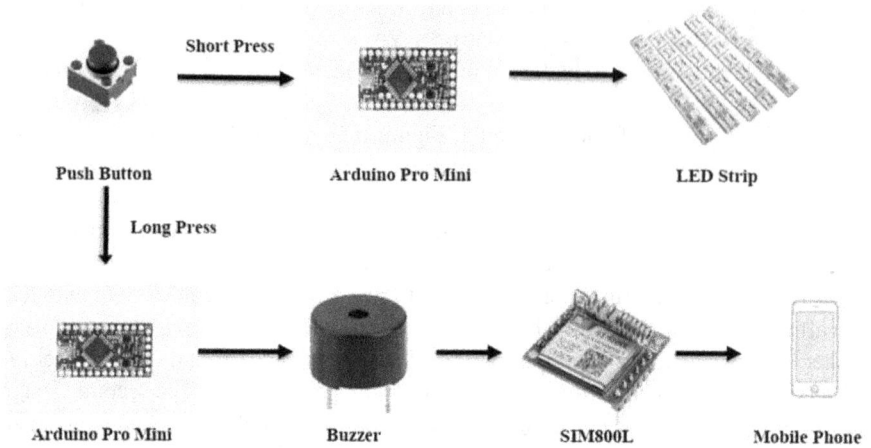

FIGURE 6.6 Process flow of emergency alarm and SMS for the proposed prototype 2.

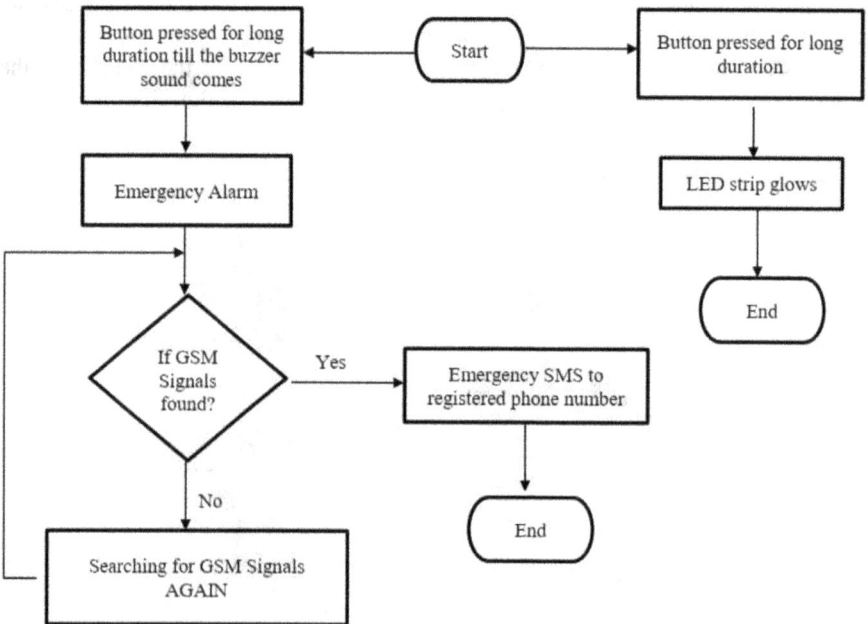

FIGURE 6.7 The flowchart for the implementation of the codes on microcontroller for the fabricated prototype 2.

Figure 6.7. C, C++ programming is used in Arduino IDE for microcontroller. Only for second design, microcontroller is used. SIM 800L GSM module is used in this circuit.

Once the switch is pressed for a longer duration, an alarm is raised, which is audible from the buzzer. Further, the GSM module part is activated. As a part of the same, the GSM module with the SIM card inserted into it locates for the strongest possible signals. Once the signals are traced, an emergency SMS is sent to the contact list as stored in the SIM card; otherwise, the GSM module keeps searching for signal strength. In the event of a very long duration spent searching the signal, the system shall be reset again so as to facilitate a fresh use of the electronic circuit.

Figures 6.6 and 6.7 together deploy this concept. The figures taken together are a step-by-step tutorial on how the coding is done so as to achieve the tasks for facilitating comfort to the user in case of an emergent situation.

6.4 FEEDBACK OF THE UMBRELLA PROTOTYPES THROUGH DESIGN THINKING APPROACH

The feedback strategy for the umbrella is kept the same as the walking stick feedback. In the feedback questionnaire, both the Primary and Secondary users are taken into consideration. *Primary users* are the elderly who shall be using the umbrella, and *Secondary users* are the ones who shall receive the SMS. The questionnaire consists of both qualitative and quantitative questions in nature. The questions along with relevant explanations are clearly marked with each question. The qualitative questions are highlighted in bold and italics. The quantitative questions are highlighted in bold and underlined as well. There are some notes added that show as Bold plus italic plus underlined. This helps in understanding the user better. However, in the whole process of interview using the questionnaire, the age of the user is asked towards the end so as to encompass the effects of the Hawthorne effect.

Further, for the Likert scale adopted in the feedback questionnaire, a scale of 1–5 is not chosen since it is not clear that some people might have injuries and so it might lead to inherent pain, which cannot be mapped on a scale of 1–5. This may lead to incorrect data collection for any question where emotions are attached. A scale of 1–10 is not chosen since this scale is chosen for more complex scales, and since this is a focused group on the elderly, a scale of 1–10 is avoided. Thus, a scale of 1–7 is included. This is adopted for the questions where a Likert scale rating is sought.

The detailed questionnaire is as follows:

Primary User (The main elderly user of the Umbrella) Here the questions pertaining to the main elderly user who uses the umbrella are laid down.

Age: .... **(It is a Quantitative question which is asked towards the end of the interview session to avoid the Hawthorne effect. If this question is posed at the beginning of the interview itself, the likelihood that the responses received by the elderly shall be biased is likely to be very high.)**

Gender:............................... **(It is a Quantitative question which is marked or used for the data analytics and ideation of correction in the prototypes after the interview session of all the interviewees is completed. Sometimes the preferences of the male and female choices differ based on Gender and so it constitutes an important point in the questionnaire.)**

Asymmetric Umbrella

Design Feedback

1. What is your opinion on the handle of the Umbrella? *(This is a Qualitative question. After the Asymmetric Umbrella is shown to the elderly user and they use it, their opinion on the umbrella, its comfort and pain points are sought as feedback.)*

2. On a scale of how much comfort do you feel with the handle of the Umbrella? **(This is a quantitative question to map the comfort level of the handle of the asymmetric umbrella).**

    (Least) 1-----|-----|-----|-----|-----|----- 7 (most)

3. What is your opinion on the support from the cuff at your hand? *(This is a Qualitative question to seek feedback on the cuff or the cloth piece that is tied in the Asymmetric Umbrella design. This helps understand whether the pan points as quoted during the initial survey before design ideations are developed are reduced or not.)*

4. On a scale of how much comfort do you feel with the support of cuff at hand? **(This question maps the comfort level of using the arm cuff support for the umbrella quantitatively.)**

    (Least) 1-----|-----|-----|-----|-----|----- 7 (most)

5. What is your opinion on design of the umbrella and any suggestions? (This is a Qualitative question that seeks the overall feedback of the developed prototype for design concept 1 of the Asymmetric Umbrella.)

Technology Feedback

6. What is your opinion about the lighting solution of the umbrella? *(This is a Qualitative question which is asked to know the user response to the new lighting solution provided in the Umbrella. The aim is to observe whether the user feels it convenient to use and how they handle it in times of need.)*

7. On a scale of 1–7, how useful is the lighting solution for you? **(The question posed above is marked on the Likert scale for achieving a Quantitative analysis of the same. This would help rate the likeability or acceptance of the lighting solution provided in the umbrellas fabricated. The least acceptance is marked as 1 and the maximum acceptance is marked as 7.)**

    (Least) 1-----|-----|-----|-----|-----|----- 7 (most)

8. How comfortable is the position of the switch and how comfortable do you feel while using it? *(As a Qualitative question, the aim of asking it is to know whether the position of the switch which is near the handle is accessible by the elderly user or not. Further, if is there any discomfort that the elderly user faces while operating the switch. If any, then that is also clearly stated.)*

9. On a scale of 1–7, how comfortable do you feel with the switch? **(The above question is marked on the Likert scale rating for a Quantitative analysis of the data. As is with the usual practice adopted in all the questionnaires referred to in the book, a scale of 1 is the least preferred and a scale of 7 is the most preferred.)**

    (Least) 1-----|-----|-----|-----|-----|----- 7 (most)

10. How useful is the emergency alarm in the umbrella? *(The question seeks the opinion and the honest feedback of the elderly user about the presence of the emergency alarm that is activated by the buzzer selection. Since, this*

seeks the opinion; it is a Qualitative question which provides relevant deeper insights into the problems, if any, in using the alarm system.)

11.   On a scale of 1–7, how useful is the emergency alarm or the buzzer placed in the umbrella? **(The above question is marked on the Likert scale rating for a Quantitative analysis of the data.)**

(Least) 1-----|-----|-----|-----|-----|----- 7 (most)

Retrofitted Umbrella Prototype 2

Design and Technology Feedback

12.   What is your opinion on the buttons and switches provided in the handle of the umbrella? *(This is a Qualitative question. Interestingly, the button or the switch for the products is different. While a simple button type switch is used for design concept 2, it provides a press button which can be operated with a long press or a short press. The long press and the short press perform different operations as is evident from the flowcharts of Figure 6.7. The short press of the button lights up the LED circuit for providing the lighting solution. Thus, the feedback is taken separately for the button or the switch provided for the prototype.)*

13.   On a scale of 1–7, how useful and comfortable are the buttons and switches provided? (The above question is marked on the Likert scale rating for a Quantitative analysis of the data.)

(Least) 1-----|-----|-----|-----|-----|----- 7 (most)

14.   What is your opinion on the SMS solution for emergency? *(As a Qualitative question, it is important to know how the elderly users feel about the SMS solution which can be used during emergency situations. The opinions can help improve the technological solutions provided.)*

15.   On a scale of 1–7, how useful is the emergency message? **(The above question is marked on the Likert scale rating for a Quantitative analysis of the data.)**

(Least) 1-----|-----|-----|-----|-----|----- 7 (most)

16.   What is your opinion on the lighting solution provided in an umbrella? *(This is a Qualitative question. The lighting solution is inherently applied to both the fabricated prototypes. The solution aims to improve the visibility of the roads during dark or power-cut moments. Through this question the elderly user place how they feel with such a solution in their umbrella as a product.)*

17.   On a scale of 1–7, how useful is the lighting solution? **(This is a Quantitative question to plot the feelings or emotions of working with a lighting solution for the walking stick. The reason for the choice of the Likert scale from 1 to 7 is the same as for previous cases.)**

(Least) 1-----|-----|-----|-----|-----|----- 7 (most)

18.   How useful is the emergency alarm in the umbrella? *(This is a Qualitative question. The opinion is sought on the alarm solution on the press of a long button as provided in the second fabricated prototype of the design concept 2.)*

19.   On a scale of 1–7, how useful is the emergency alarm (buzzer)? **(This is a Quantitative question. The aim is to map the Likert scale rating of the qualitative inputs sought for the above question so as to map it for statistical treatment. The reason for the choice of the Likert scale from 1 to 7 is the same as for previous cases.)**

(Least) 1-----|-----|-----|-----|-----|----- 7 (most)

20.   Any suggestions? (The elderly users are asked to make any additional suggestions about the umbrella, if they have any. This question is also treated qualitatively.)

Secondary User (The person who takes care of the elderly umbrella user or who specifically receives the emergency SMS through the SIM inserted in the Umbrella. On the secondary end, it can be a single user or multiple users, depending on how many user contacts should be sent the SMS details which can be programmed in the microcontroller programming it. Usually not more than five contacts are addressed or a priority contact list is provided to ensure that such emergency details are sent to them. Further, the role of the secondary users comes into the picture for the second fabricated prototype, that is, the design concept 2.)

21.   What is your opinion on the alarm and SMS solution for emergency? *(This is a Qualitative question. The inputs sought from the secondary users help improve the concept of emergency SMS reception of the elderly amidst an emergency-based situation. The secondary users can advise on how to improve the system for better comprehensibility or reduce complexity for the elderly user.)*

22.   On a scale how useful are emergency alarm and SMS? **(This is a Quantitative question. The aim is to map the Likert scale rating of the qualitative inputs sought for the above question so as to map it for statistical treatment. The reason for the choice of the Likert scale from 1 to 7 is the same as for previous cases.)**

(Least) 1-----|-----|-----|-----|-----|----- 7 (most)

23.   Any Suggestions? *(The secondary users of the elderly, who use the prototyped umbrella, are asked to make any additional suggestions about the umbrella, if they have any. This question is also treated qualitatively.)*

6.5   DISCUSSION AND CONCLUSION

While taking the feedback, users are categorised into two types.

1. *Primary user* – They constitute the main user or the elderly using the umbrella.

   From the primary user, the feedback is collected twice.

   I.   For assessing the design of the umbrellas. Design feedback is taken from 27 users. This feedback and the questionnaire designed for this purpose is already deliberated in detail in the previous chapter.
   II. For assessing the technology intervention of umbrellas. Technology feedback is taken from 19 users.

2. *Secondary user* – The person who is the caretaker of primary user. Secondary user actually receives the emergency SMS. Secondary user feedback is taken from 19 people.

Feedback given by the users is plotted on a pie chart. Likert scale is used to note down the response of the users and for projection of the quantitative data gathered from the users. Their response is also noted in terms of some words and numbers. These are as mentioned below:

- Very Good – 7
- Good – 6
- Above Average – 5
- Average – 4
- Below Average – 3
- Not Bad – 2
- Bad – 1
- No Response – 0

It was interesting to note that after 12 users, the responses started reaching a saturation level. The feedback is analysed and explained using the pie charts. Here the pie charts are presented one by one for three designs of the umbrellas. Among the people from whom the feedback is taken, some people state it to be good, some state average, etc. So, the responses are mentioned in percentages.

Figures 6.8 to 6.12 show the feedback of the Asymmetric umbrella with the technology intervention and Figures 6.13 to 6.17 show the feedback of the conventional umbrella with primary and secondary user feedback. The qualitative responses for the feedback conducted on various design and fabricated prototype aspects were found to be in tandem with the idea behind the two ideated design concepts. Hence, they are not discussed here, as it is included in the ideation stage itself.

Out of all the elderly who participated in the feedback survey, 95% of the users gave positive responses to the handle. They found it more comfortable and

1ST DESIGN HANDLE

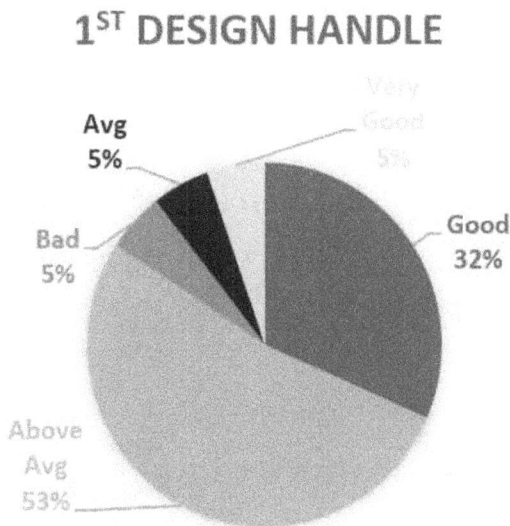

FIGURE 6.8 Feedback of 1st design handle.

ergonomic, and it exerted less pressure on the shoulder as compared to conventional umbrellas. This is shown in Figure 6.8.

Seventy-four per cent of the users responded positively to the handcuff provided in the umbrella. Twenty-six per cent of the users did not respond regarding the cuff. Some of them said they can use an umbrella even without the cuff. Twenty-six per cent people were initially hesitant to use the handcuff; however, once they were explained the necessity of it, the feedback was good. This is shown in Figure 6.9.

All of the elderly users who participated in the feedback survey gave positive responses to the lighting solution. While 42% said it was good, the remaining 58% of the elderly found it above average. This plot of pie diagram is shown in Figure 6.10.

All of the elderly users were positive about the switch which was provided in the umbrella. The switch was found to be in an appropriate position for the elderly to operate on, and it was found to be easily operable. These data are recorded in the plot of Figure 6.11.

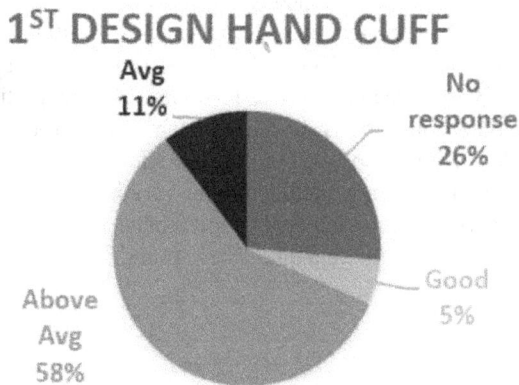

1ST DESIGN HAND CUFF

Avg 11%

No response 26%

Above Avg 58%

Good 5%

FIGURE 6.9   Feedback of 1st design handcuff.

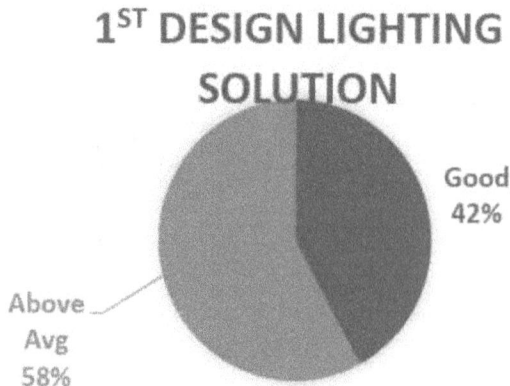

1ST DESIGN LIGHTING SOLUTION

Good 42%

Above Avg 58%

FIGURE 6.10   Feedback of 1st design lighting solution.

1ˢᵀ DESIGN SWITCH

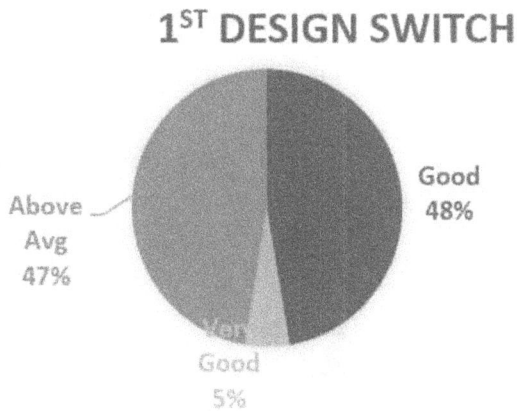

Above
Avg
47%

Good
48%

Good
5%

FIGURE 6.11 Feedback of 1st design switch.

1ˢᵀ DESIGN EMERGENCY ALARM

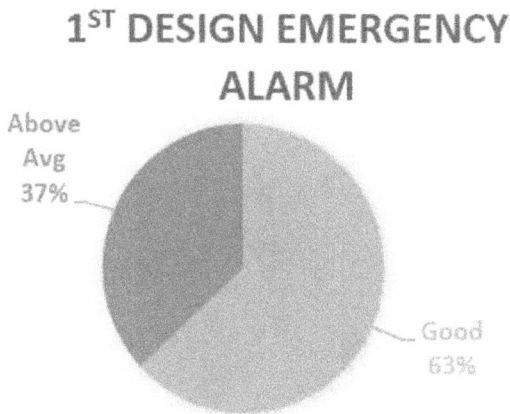

Above
Avg
37%

Good
63%

FIGURE 6.12 Feedback of 1st design alarm.

The emergency alarm provided in umbrella design 1, that is, the prototype of the Asymmetric umbrella was appreciated by all the elderly users of the survey. Keeping in mind the age constraint of 60 years and above, an alarm solution would attract aid from people nearby immediately. This is plotted in the pie diagram of Figure 6.12.

For the second design, the elderly participants of the feedback survey approved the design of the switch and the buttons which are used. While 44% of the users find it above average, 56% said that it is average in nature. This is plotted in Figure 6.13.

The emergency SMS sent from the umbrella was much appreciated by all the elderly participants of the feedback survey. Twenty-one per cent of the users found it to be very good, and around 58% found it to be good enough. The main reason for the acceptance was the fact that in emergent situations, it becomes very important for elderly users to connect to their near and dear ones. This plot is shown in Figure 6.14.

2ND DESIGN SWITCH & BUTTON

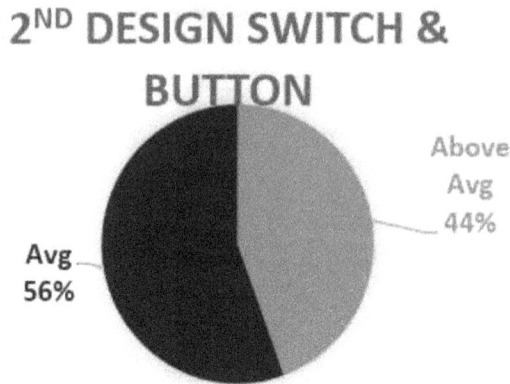

Above
Avg
44%

Avg
56%

FIGURE 6.13   Feedback of 2nd design switch and button.

2ND DESIGN EMERGENCY SMS

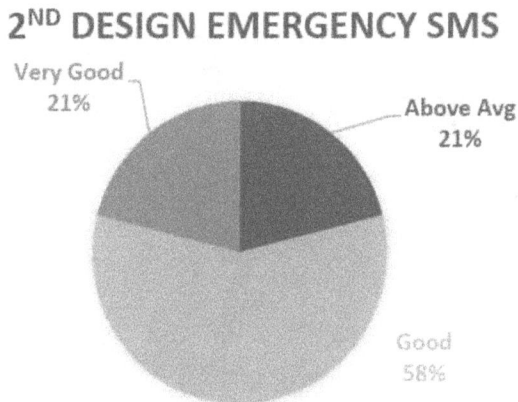

Very Good
21%

Above Avg
21%

Good
58%

FIGURE 6.14   Feedback of 2nd design emergency SMS.

95% of the users gave positive response to the lighting solution provided in the umbrella. It was found to be in tandem with the Affordable principle of the Design. The plot is shown in Figure 6.15.

Ninety-five per cent of the users found the emergency alarm provided in the umbrella to be extremely useful. While 53% found it to be good, most of the elderly users approved the use for emergent purposes. This is shown in Figure 6.16.

Among the 19 secondary users who were surveyed for the efficacy of the SMS reception on their respective mobiles, 58% found it to be good and all of the secondary users approved of the concept. This is to facilitate the fact that the elderly user can be tracked in case of an emergency situation while travelling alone. This is shown in Figure 6.17.

2ND DESIGN LIGHTING SOLUTION

Good 45%

Above Avg 50%

Below Avg 5%

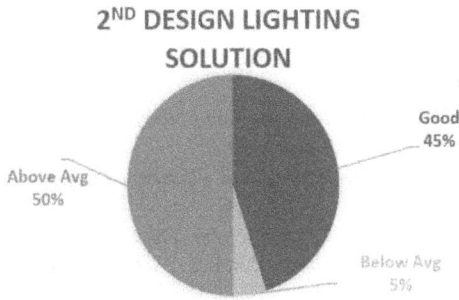

FIGURE 6.15 Feedback of 2nd design lighting solution.

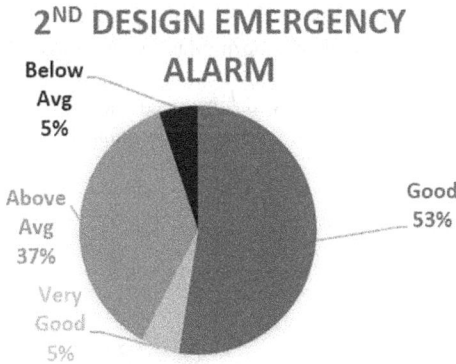

2ND DESIGN EMERGENCY ALARM

Below Avg 5%

Above Avg 37%

Good 53%

Very Good 5%

FIGURE 6.16 Feedback of 2nd design emergency alarm.

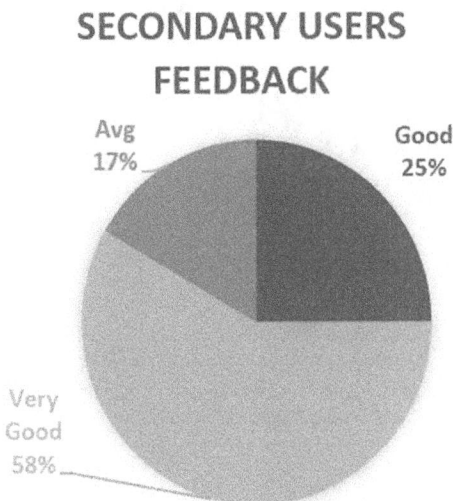

SECONDARY USERS FEEDBACK

Avg 17%

Good 25%

Very Good 58%

FIGURE 6.17 Feedback of 2nd design emergency SMS secondary users.

REFERENCES

1. Nigel Cross, *Design Thinking: Understanding How Designers Think and Work*, Berg Publishers, 2011.
2. Tim Brown, Clayton M. Christensen, Indra Nooyi, Vijay Govindarajan, "On Design Thinking", *Harvard Business Review*, 2020.
3. Bill Burnett, Dave Evans, *Designing Your Life: How to Build a Well-lived, Joyful Life*, Knopf, 2016.
4. Michael Lewrick, Patrick Link, Larry Leifer, *The Design Thinking Playbook: Mindful Digital Transformation of Teams, Products, Services Businesses and Ecosystems*, Wiley, 2018.
5. J. Robert Rossman, Mathew D. Duerden, B. Joseph Pine II, *Design Experiences*, Columbia Business School Publishing, 2019.
6. https://lastminuteengineers.com/sim800l-gsm-module-arduino-tutorial/
7. https://robu.in/product/tp4056-1a-li-ion-lithium-battery-charging-module-with-current-protection-mini-usb/
8. https://www.arduino.cc/en/pmwiki.php?n=Main/ArduinoBoardProMini
9. N. Raghu, V. N. Trupthi, N. Krishnamurthy, K. Rasagnya, D. A. Darshan, "Effects of Electromagnetic Field on Patients with Implanted Pacemakers", *International Journal of Advance Research and Innovative Ideas in Education*, Vol. 1, No. 5, pp. 38–41, 2016.

7 Technology Inputs to Design for Products

People often misconceive the competing field of Fine Arts with Design. In European Academic institutions, Fine Art is said to be art developed with the main aim to focus on aesthetics or beauty, distinguishing it from decorative art or applied art [1]. As an example, someone may easily draw a line or simply sketch a line on a piece of paper. This shall be treated as Fine Art, provided it gets quantified or provides a meaning on why a line is sketched. This quantification or adding meaning to the sketch lays the first step of the design.

Design imparts values which graduate as tangible outputs. These outputs could be in the form of product, interaction, space, UI/UX (User Interface or User Experience), etc. Interestingly any design remains unwarranted without the participation of the user, which brings forth the field of Ergonomics to be closely associated with Design [2, 3].

As is well explained in [2], Engineering and Technology aid to improve the products or process from viewpoint of the product only. Psychology refers to the mind mapping and study of behaviour. Human factors and ergonomics are concerned with adapting the products to the people, based upon their physiological and psychological capacities and limitations.

Keeping all this in purview, this book is written to understand the journey from design to technology. The concepts of the design thinking approach are well adapted to start the design process from scratch and the fabrication issues, methodology and technology intervention have also been well articulated as referred from [4–7].

7.1 TECHNOLOGY IN MARKET

It is interesting to note that the recent era has flooded the market with a plethora of technologies. Technology means the application of scientific knowledge for practical purposes. Technologies are meant to ease work, force and efforts and also reduce cognitive load. The role of technology is always to ensure that users should be able to use it easily. This is based on the Aesthetic-Usability Effect of the Design Principle. If the users are unable to address the use of the technology, then it needs significant introspection. It is quite obvious that since the human race and ethnicity are different across the world, the same technology cannot be replicated without undergoing design intervention.

As a small example, the way we hold a spoon for eating food cannot be the same throughout the world. The grip to hold the spoon can differ from person to person. This brings the ergonomics to understand the usage of spoons across different parts of the world.

Similarly, in the book, the target participants have been clearly articulated from the beginning, the users being the Indian elderly. Looking at the average literacy of

DOI: 10.1201/9781003414957-7

the Indian elderly, none of the proposed technology solutions has been kept complex, allowing ease of operation for the average Indian elderly encompassing both males and females. The technology should adhere to the accessibility principle so that maximum number of people should be able to use the product without much difficulty.

This proposal lays emphasis on the man–machine interaction and the pyramid. Figure 7.1 helps identify this situation. It is interesting to note that whereas man lies on top of the pyramid, technology or machine and the environment or the context form the base of the pyramid. Thus, while humanizing technology is concerned, the environment or context plays an important role.

It should never be forgotten that technology is supposed to aid human beings. Looking at technology from a system perspective also gives an input that if technology has a reliability of 0.8, human has a reliability of 0.8, then the human machine or technology together has a reliability of $0.8 \times 0.8 = 0.64$. Thus, in the presented context, since the target group is the average Indian elderly, the technology has been kept as simple as possible so that the system reliability can be improved. This factor should always be kept in mind while deriving any innovative product from new technologies.

The figure also clearly shows that the technology always remains at the back end. For example, when we buy a new product, no one opens the product to check its components; rather, its data sheet is sufficient to provide technology inputs and what matters is its look, ergonomics and aesthetics. This is where the role of designers comes to the prima facie.

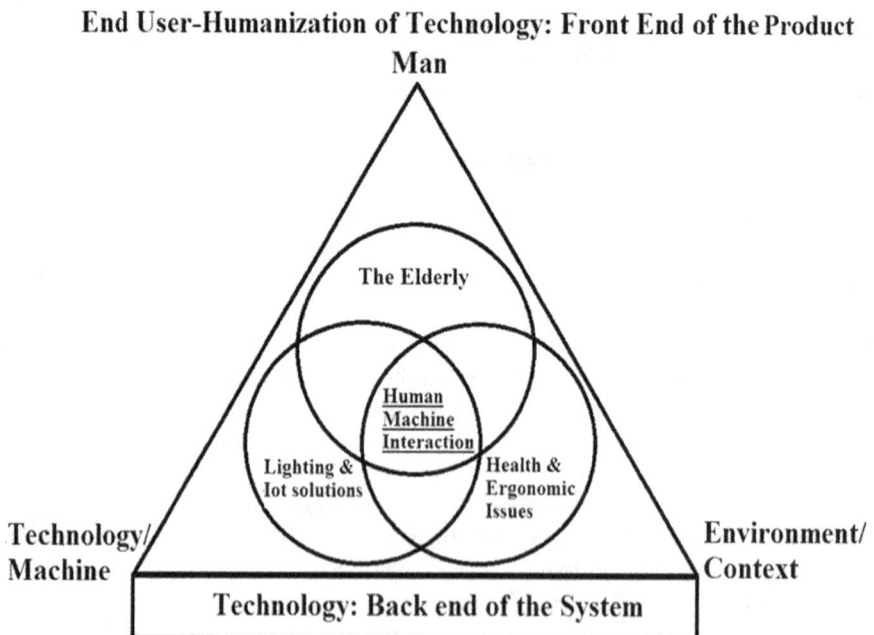

End User-Humanization of Technology: Front End of the Product

Man

The Elderly

Human Machine Interaction

Lighting & Iot solutions

Health & Ergonomic Issues

Technology/ Machine

Environment/ Context

Technology: Back end of the System

FIGURE 7.1   Man–machine pyramid with technology as the back end.

7.2 DESIGN INPUTS TO TECHNOLOGY

Since technology lies at the backend of any system-level perspective, design inputs to improve the system performance play a crucial role. This can also be explained in Figure 7.2. The detailed methodology followed for the re-design of the walking stick and umbrella for the elderly is aligned with nature.

The initial steps of Figure 7.2, till data analysis and re-design occupy the maximum time. The Initial Data collection and Analysis are based on Direct Observation and Activity Analysis with Questionnaires and interviews and subsequently by means of chunking, empathy maps and mapping of pain points. Through the data collected by these processes, the initial design concepts for both products are proposed.

The technology mapping through process flow and journey maps undergoes several iterations so as to retrofit the technology in the designed concepts and fabricated prototypes. After the testing and validation along with user feedback, the product can be launched into the market.

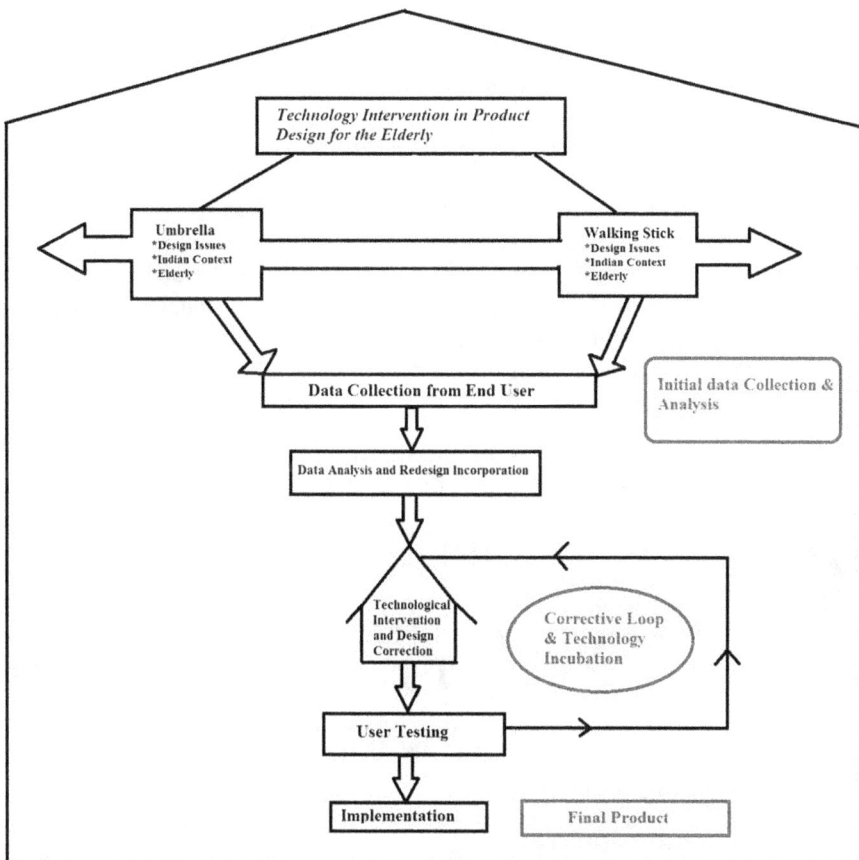

FIGURE 7.2 Methodology for the re-design of the products and technology intervention.

Whereas engineering and technology solutions are important, the role of designers to make the system more plausible cannot be overruled. Technology and design together contribute to the cost-benefit factor, where a product is adopted only if its benefits are equal to or greater than the costs involved.

While technology provides various solutions, it is the designers who entertain entry points in the usage of the design with minimal barriers and progressive lures.

7.3 TECHNOLOGISTS VERSUS DESIGNERS

There is always a trade-off between sustainable and universal products. The more a product tends to be universal, the less sustainable it becomes and vice versa. In view of the same, the technology as a backend aids the designers to add universal nature to a product than make it just sustainable. For example, a mobile phone was initially launched only for voice calls and to send SMS or short message services. Later on, with the advent of technology, rolling internet on mobile phones changed the perspective of calling. People now use mobile phones for video calls, watching movies and web series, placing orders to various apps, monitoring health issues, etc.

While the initial goal of the design was voice call, which still remains; technology at the back end, adding to various resources, has made a cell phone more universal in nature. For any product to sustain in today's world, contribution of technology and design has to go hand in hand.

For the complete case of Product design for the elderly with emphasis on walking stick and umbrella re-design, two important constraints have always been pivotal. This is also shown in Figure 7.3. First, the proposed design and technology solution should cater to both, socio-economic and techno-commercial solutions. Elderly from all socio-economic realms should be able to afford the solutions and also comfortably use the technology inputs as solutions to the products. Second, inspired by IBM innovation [8], it is very important that the product and technology should be user-centric. User-centric designs cater to ergonomic issues and are more acceptable in nature than technology-centric designs.

(a)   (b)

FIGURE 7.3 Two important constraints of walking stick and umbrella for the elderly. (a) Socio-economic and techno-commercial solution. (b) IBM Human Innovation Center inspired solutions.

Designers and technologists also need to understand the flexibility–usability trade-off. The more a product or system becomes flexible, the less usable it becomes. This is also analogous to Hick's law, that more the options are given to a user, more the time it takes to reach a decision. Thus, technology inputs in this case have been constrained so as not to add any further cognitive load on the elderly.

The designer and technologist relationship is well balanced by the hierarchy of needs of any design and usage of the technology associated. Any product works on the principle of satisficing. Any technology cannot satisfy every user, but if it caters to the requirements of the maximum number of users, then the technology can be rolled out. For this purpose, feedback for both the design and technology from primary and secondary users is extremely important. Thus, designers and technologists need to work at the same pace and scale in order to contribute to a product to cater to the maximum number of masses.

REFERENCES

1. "Aesthetic Judgement", *The Stanford Encyclopedia of Philosophy*, 2010.
2. Neville Stanton, *Human Factors in Consumer Products*, Taylor & Francis, 2003.
3. Prabir Mukhopadhyay, *Ergonomics for the Layman*, CRC Press, Taylor & Francis, 2019.
4. Siddhu Kulbir Singh, *Methodology of Research in Education*, New Delhi, Sterling Publisher, 1992.
5. S. P. Sukhia, P. V. Mehrotra, *Elements of Educational Research*, Allied Publisher Private Limited, New Delhi, 1983.
6. Martyn Denscombe, *The Good Research Guide*, New Delhi, Viva Books Private Limited, 1999.
7. William Lidwell, Kritina Holden, Jill Butler, *Universal Principles of Design*, Rockport Publications, 2003.
8. https://www.ibm.com/marketing/hci

Index

Pages in *italics* refer figures and pages in **bold** refer tables.

For Product Safety Concerns and Information please contact our EU
representative GPSR@taylorandfrancis.com
Taylor & Francis Verlag GmbH, Kaufingerstraße 24, 80331 München, Germany